TOWN HALL POWER
or
WHITEHALL PAWN?

Town Hall Power or Whitehall Pawn?

Local autonomy=local revenue · Misuse of welfare as power base
Deficiencies of research · Ineffectiveness of social work
Economics of politics and bureaucracy · Saving local government

Introduction by the Seminar Chairman:
ARTHUR SELDON

DAVID KING

DIGBY ANDERSON • **W. W. SHARROCK**

JUNE LAIT • **DAVID MARSLAND**

KEITH HARTLEY • **GEORGE W. JONES**

With Commentaries by
Linda Whetstone · **Merrie Cave** · **Anthony Bays**

SIR JOHN GRUGEON

Session Chairman:
LORD HARRIS OF HIGH CROSS

Published by
The Institute of Economic Affairs
1980

First published in December 1980 by
THE INSTITUTE OF ECONOMIC AFFAIRS

ISSN 0305-814X
ISBN 0-255 36135-1

Printed in England by

Goron Pro-Print Co. Ltd., Lancing, West Sussex

Set in Monotype Times Roman 11 on 12 point

Contents

Preface

The IEA *Readings* have been devised to refine the market in economic thinking by presenting varying approaches to a single theme. They are intended primarily for teachers and students of economics but are edited to help non-economists who want to know what light economics can shed on their activities.

Readings 25 is based on a Seminar in June 1980 organised to examine the economic implications of the relationship between national and local government.

The opening paper by Dr David King analysed the effects on local autonomy of the dependence on Whitehall for much of its finance. The next four papers, by sociologists associated with the new Social Affairs Unit, Dr Digby Anderson, its Director, David Marsland, Deputy Director, June Lait and Wesley Sharrock, Consultants, discussed how far local government was competent to provide welfare services, which consume most of its expenditure.

The economics of politics and bureaucracy in local government was then examined by Dr Keith Hartley, and the Seminar was rounded off by Professor George Jones on the ways to avoid over-dependence on national government and thus how to save the autonomy of local government.

The lunchtime address was delivered by Sir John Grugeon, Chairman of the Policy Committee of the Association of County Councils, who discussed the theme of the Seminar as seen by a leader in local government.

The Seminar was made especially timely by the recent tendency of local government to spend beyond its resources and the attempts made by the new Government in Whitehall to control local over-expenditure through its supply of finance in the Rate Support Grant. There is no limit to the services that local government can claim its populations would desire, or that it thinks those 'under-privileged' should have, but there is a limit to the revenue they can collect locally in rates, not least because people running businesses and even householders can ultimately move out of districts or counties which levy local taxes they consider unacceptable. Yet the commonly discussed escape from this limit—that the cost of local services shall be shifted from local rates to national taxes—is no lasting solution. If there is general revulsion against taxes, it applies to national tax as well as to local rates. Rate-payers will no more like higher central taxes to pay for local services than they will higher local rates. The supposed escape, moreover, raises the dilemma that if local authorities come to depend increasingly on central government to finance local services, they will have to accept increasing influence and ultimately regulation by central government, and their status as local governments will be increasingly undermined.

This was the flaw in the report of the Layfield Committee that reported on local government financing in 1976. Its recommendation of a local income tax to replace the sagging revenue from the rates was no solution at all. To substitute one tax for another may remove anomalies and possibly tap some additional revenue, but it does not remove the underlying difficulty that the populace simply does not wish to pay higher taxes of any kind, local, regional or national, direct or indirect, on earnings or purchases, on income or wealth, excise or customs. Nor is this surprising, since taxes are raised to finance services in which the populace has little or no say, and which it could pay for more satisfyingly in the market.

Attention must, therefore, increasingly be given to a neglected source of local revenue to which objections will not be raised since it is not a tax and which has the formidable advantage that it would strengthen rather than undermine local autonomy. That solution is the one that would suggest itself most naturally to the economist. Some of the services provided by local government can be provided only by government, although there is room for disagreement on how far they must be provided locally or nationally (or regionally). But many, or possibly most, of the services, in value or quantity, do not have to be supplied by local government. They are often a relic from the last 50 or 100 years in which they were established possibly with plausible reason that no longer applies in the last two decades of the 20th century. It is here that there is scope for both raising revenue by charging, or by charging nearer to costs, or by reducing expenditure by hiving them off to private enterprise in the market. And although it may be more spectacular for government to decide which services shall be transferred *holus-bolus* from local government to the market, the more sensitive and more efficient way would be to raise charges (in the form of fees, fares, dues, rents, etc.) and so allow consumers themselves to decide whether they wish to continue the services to be supplied by local government or by competitive private suppliers.

These issues are discussed in the Seminar papers by the two economists, four sociologists and the political scientist, and they supply material for a fundamental reconsideration of the financing of local government and its significance for local government autonomy.

September 1980 ARTHUR SELDON

Introductory Remarks by the Chairman

ARTHUR SELDON

The Author

ARTHUR SELDON was born in London in 1916, educated at Raine's Foundation School and graduated from the London School of Economics. He was a Tutor in Economics for the London University Commerce Degree Bureau, 1946-56, and a Staff Examiner in Economics to the University, 1956-66.

After editing a retail journal and advising in the brewing industry, he joined the Institute and wrote its first *Paper* in 1957 (on pensions), its early reports on advertising, hire purchase and welfare with Ralph Harris, and several later *Papers*. He is co-author (with Ralph Harris) of *Pricing or Taxing?* (1976), *Not from Benevolence . . .* (1977), and *Over-ruled on Welfare* (1979); and he contributed 'Change by Degree or by Convulsion' to *The Coming Confrontation* (1978).

He wrote *The Great Pensions Swindle* (Tom Stacey, 1970), and compiled *Everyman's Dictionary of Economics* with the late F. G. Pennance (J. M. Dent, Second Edition, 1975). He wrote *Charge* for Maurice Temple Smith (1977), and contributed an essay, 'Individual Liberty, Public Goods and Representative Democracy', to the Hayek Festschrift (*ORDO*, 1979). He has been a contributor to the *Economist,* the *Financial Times*, *The Times* and the *Daily Telegraph.*

Welcome to this, the 14th IEA Seminar of a series started in 1972. They are now something of a bi-annual event. In them we attempt to assemble both economists and men of the world—the practitioners—and on occasions we mix our 'own' economists with other scholars. Today we have a possé of 'our' sociologists from whom you may hear opinions you may not expect. We have therefore assembled a number of university teachers and local councillors and members of their staffs. We even have councillors from the 'interesting' boroughs of Lambeth and Camden and a number of academics who are also councillors. We also have a number of Whitehall bureaucrats, if I may call them that, writers and, most important of all, people from industry who create our incomes, our salaries and our pensions. We should therefore have a varied and stimulating day, not least because the subject of local government has been increasingly topical for some time.

How far can *local* government be independent if a large amount of its income is supplied by *central* government? And what does it require to make itself truly the *de facto* local *self*-government? I will not pretend that I have no views on this subject, although I shall be the impartial arbiter that chairmen ought to be. But I think I must say that government as a whole is in increasing disrepute, rather like a declining industry—we do not think much of its services or products. More and more economists have been arguing that we want less and less government; yet for decades past we have had more and more government. It is sad to have to say that in the past year under the new Government we have had even more government. We now have more government than we had last year, and that under a Government that said we would have less government.

Symptoms of the rejection of government

The rejection of government is seen in tax evasion and tax avoidance and in the increasing number of irate-payers, or high rate-payers. Our task is going to be to discover ways and means of cutting down government to the limits required by public opinion. I think we are going to have a very interesting next 10 years. The argument will be increasingly intense in academia, in the Press and, I hope, not least in the House of Lords. We shall have to go back to first principles. We shall have to ask what government, local or central, should do. Should it supply only public goods? What should it *not* do? What should it leave to firms and individuals to provide for themselves by exchange in the market? How should government be disciplined? Should local government supply libraries, abbatoirs, housing, airports? And, if so, why? Should government have its own training colleges for office staffs when we have discovered that private colleges can train office staffs at a third to a half of the cost of government colleges of further education? Who should pay for local services? How should they pay?—and how much? These are the questions our platform of speakers will discuss. I hope our deliberations and exchanges will be scholarly, crisp in language, strong in their conclusions for policy, and where possible new.

We start with Dr King. His subject is one that is really our major theme: How far can local government continue to be local and autonomous if a large amount of its income is derived from other sources? A mark of David King's authority to speak on this subject is that he was on the staff of the Kilbrandon Commission. The four tables in his paper have been circulated, and we have added a table on employment in the public and private sectors and local government separately (p. 149).

PART I
IS LOCAL AUTONOMY POSSIBLE WITHOUT LOCAL REVENUE?

1. Local Government Finance and Local Government Autonomy

DAVID N. KING
University of Stirling

The Author

DAVID N. KING: Lecturer in Economics at the University of Stirling since 1978. Graduated from Magdalen College, Oxford, and obtained a doctorate at the University of York. Was consultant economist to the Royal Commission on the Constitution, 1971-72, and wrote one of its Research Papers. Taught economics at Winchester College, 1972-78. For the IEA he wrote, with Alan K. Maynard, *Rates or Prices?* (Hobart Paper 54, 1972).

Commentary Author

LINDA WHETSTONE: Former parish councillor. Educated at St. George's School, Ascot; Rockford College, Illinois; American University, Washington and the London School of Economics. For the IEA she has written *The Marketing of Milk* (Research Monograph 21, 1970); with Henry Smith, *A Market for Animal Semen?* (Research Monograph 29, 1972); and a study of egg prices in *Essays in the Theory and Practice of Pricing* (Readings in Political Economy 3, 1967).

I. INTRODUCTION

For the past century, local authorities in Great Britain have been becoming increasingly dependent on grants from the central government to finance their current expenditure. This trend has led to widespread concern amongst those who believe that growing central government finance tends to produce increasing central control and so to reduce local autonomy. The most favoured solution to this problem is to give local authorities additional tax powers. Another approach, which should be seen as a complementary solution rather than a substitute, is to reduce the cost of the services local authorities provide from rate and grant receipts, either by allowing them to make charges or by transferring provision to private enterprise or central government.

In this paper I will consider this debate in more detail. I will also mention another approach to easing the problem, that of devising a grant system which permits as much local discretion as possible. I would not claim that this solution is the most effective: on the contrary, it is perhaps the least effective. But it deserves particular consideration at present, for the Government is changing the grant system in the 1980 Act.[1]

II. TRENDS IN LOCAL REVENUE AND EXPENDITURE

The two chief sources of revenue on local authority current accounts are rates and government grants. The authorities also obtain a sizeable income from other sources, principally rent, trading surpluses, interest, fees and charges. Around the turn of the century, grants amounted to about 15 per cent of the total and rates to over half, but the growing scale of local education and housing expenditure in the early years of this century meant that by 1923-24 the grant share had risen to 22 per cent and the rates share had fallen to about 42 per cent. The trend since then is shown in Table I. It will be seen

[1] *Local Government Planning and Land (No. 2),* Bill 128, January 1980. [Now the *Local Government Planning and Land Act*, Chap. 65, 1980.—ED.]

TABLE I

SOURCES OF REVENUE INCOME FOR LOCAL AUTHORITIES IN ENGLAND AND WALES, SELECTED YEARS, 1923-24 TO 1977-78

	Rates £m.	*Government Grants £m.*	*Miscellaneous £m.*	*Total £m.*	*Rates %*	*Government Grants %*	*Miscellaneous %*
1923-24	143·3	75·2	122·8	341·3	42·0	22·0	36·0
1933-34	148·6	121·6	176·4	446·6	33·3	27·2	39·5
1943-44	204·1	228·4	306·5	739·0	27·6	30·9	41·5
1953-54	392·5	414·2	369·6	1,176·3	33·4	35·2	31·4
1963-64	923·1	1,022·4	826·7	2,772·2	33·3	36·9	29·8
1973-74	2,414·6	3,897·3	3,518·3	9,830·2	24·6	39·6	35·8
1974-75	2,927·3	5,651·8	3,262·7	11,841·8	24·7	47·7	27·6
1975-76	3,795·7	7,666·1	4,000·8	15,462·6	24·5	49·6	25·9
1976-77	4,151·0	8,639·8	4,638·6	17,429·4	23·8	49·6	26·6
1977-78	4,686·7	9,138·3	5,197·3	19,022·4	24·6	48·0	27·3

Sources: Local Government Financial Statistics 1973-74, HMSO, 1975, p. 8, Table 3; *Annual Abstract of Statistics 1980,* HMSO, 1980, Table 16.43.

that by 1977-78 rates were contributing less than a quarter of current income and grants about one-half.

The falling share of rates is in no way a result of local authorities making less use of their single tax. Rates in England and Wales rose as a percentage of GNP from 3·4 per cent in 1922-23 to 3·8 per cent in 1975-76, and the decline to 3·6 per cent since then is at least partly attributable to central government pressure, for central governments are now very concerned to keep poundages down since they affect the retail price index. The problem with rates is that they have risen only slightly in proportion to GNP whilst local expenditure has risen very rapidly. This increase is shown in Table II, which demonstrates that local expenditure rose most rapidly in relation to GNP during the 1960s.

Factors in the expansion of local expenditure

Two factors have been causing the increase in local authority expenditure. First, there has been an expansion in the real scale and

TABLE II

LOCAL GOVERNMENT REVENUE EXPENDITURE IN PARTS OF GREAT BRITAIN AS A PERCENTAGE OF GNP AT FACTOR COST: SELECTED YEARS, 1923-24 TO 1976-77

(1) Year	(2) Expenditure in England & Wales (£m.)	(3) Expenditure in Scotland (£m.)	(4) Total Expenditure in Great Britain (£m.)	(5) GNP at factor cost (£m.)	(6) (4) as a percentage of (5) (%)
1923-24	343	47	390	4,201	9·3
1933-34	433	58	491	4,105	12·0
1943-44	698	87	785	8,530	9·2
1953-54	1,128	142	1,270	15,280	8·3
1963-64	2,668	308	2,976	27,692	10·7
1973-74	9,733	1,001	10,734	67,935	15·8
1974-75	11,734	1,334	13,068	80,188	16·3
1975-76	14,961	1,519	16,480	98,590	16·7
1976-77	17,133	1,917	19,050	115,028	16·6

Sources: Local Government Financial Statistics, op. cit., p. 8, Table 3; *Annual Abstract of Statistics, op. cit.*, Tables 16.41 and 16.51; *Local Financial Returns Scotland 1968-69*, HMSO, 1971, Table VIII; *The British Economy Key Statistics 1900-1970*, published for the London and Cambridge Economic Service by Times Newspapers, Table A; and *National Income and Expenditure 1979*, the Blue Book, HMSO, 1979, Table 1.1. The 1943-44 GNP figure was estimated from A. T. Peacock and J. Wiseman, *The Growth of Public Expenditure in the United Kingdom*, George Allen & Unwin, 2nd edition, 1967, p. 154.

level of their services. Increases in the school-leaving age, in the number of students entering higher education and requiring grants, and in staff-pupil ratios, for instance, have raised the real value of education budgets, whilst increases in car ownership, the crime rate and the proportion of old people in the population have led to increased expenditure on the police, roads and welfare services; and higher capital spending on all these services and others (especially housing) has raised local debt and interest payments. On the other hand, local authorities have suffered a consequence of the 'relative price effect'. It is argued that local services are typically labour-intensive so that productivity rises more slowly than in the more capital-intensive industrial sector of the economy. When productivity in the industrial sector improves, it is likely that employees will secure a rise in real wages. This will lead to a movement of labour away from local government unless local authorities give comparable increases in real wages. As a result, local authorities find that the real cost of providing a given standard of service rises over time.

III. GRANTS AND LOCAL AUTONOMY

The belief that he who pays the piper calls the tune has led many to conclude that the increasing share of local expenditure met by grants must lead to increased central control, though the hypothesis is extremely hard to prove. Over 80 per cent of the grants received by local authorities are general grants which may be spent on any service, so central control is not closely tied to the grant payments. The main control methods were succinctly noted in 1967 by the Committee on the Management of Local Government which declared that

> 'the freedom of local authorities is circumscribed by a strict application of the principle of *ultra vires,* which means that they must be able to point to a specific statutory sanction for every act. It is further limited by ministerial regulations, the need to obtain frequent consent from government departments to proposed courses of action, and a rigorous control over capital expenditure'.[2]

In 1971 the then Secretary of State for the Environment, Mr Peter Walker, stated that his department alone had more than 1,000 statutory controls over the expenditure of local government.[3]

Most of these controls have been established in the present century, concurrently with the rise in central finance of local spending. It is possible, however, that the increased control has emerged from a desire for more nationwide uniformity and is independent of financial sources. Certainly this was the view of the late John P. Mackintosh, MP, who wrote that:

> 'this desire for uniformity, this fear of the political consequence of

---

[2] Committee on the Management of Local Government, *Management of Local Government,* Volume 1: *Report of the Committee*, HMSO, 1967, p. 11.

[3] Reported in *The Times,* 23 March 1971.

> diversity and of an element of devolution of actual power is the major explanation for the myriad controls imposed on local government'.[4]

Alternatively, the controls may have grown up as a result of a genuine altruistic belief in their desirability by government departments. Submitting evidence to the Royal Commission on Local Government in England, the Ministry of Housing and Local Government said it 'cannot trust a multitude of small local authorities to spend money wisely'.[5]

Local government commissions recommend financial autonomy

Nevertheless, it is widely believed that increases in central finance of local government are likely to lead to increases in central control. The view that autonomy in action is related to autonomy over finance has been supported by virtually all Commissions on local government. The Committee on the Management of Local Government said that its

> 'researches abroad have brought out very clearly the close connection between freedom from central control and the possession of adequate local taxable resources'.[6]

The Royal Commission on Local Government in England noted that

> 'examination of foreign practice confirms the close connection between local self-government and the possession of adequate taxation powers',

and felt it was 'clear that local government needs new sources of income'.[7] The Royal Commission on Local Government in Scotland referred to the 'central control that inevitably goes with the grant', and stated:

> 'it is absolutely fundamental that local authorities should be enabled to raise a much higher proportion of the money they spend'.[8]

---

[4] J. P. Mackintosh, *The Devolution of Power,* Penguin Books, 1968, p. 100.

[5] Lawrence Welsh (ed.), *Royal Commission on Local Government, an edited reprint of evidence published in the Local Government Chronicle,* Charles Knight, 1967, p. 34.

[6] Committee on the Management of Local Government, *op. cit.,* p. 13.

[7] Royal Commission on Local Government in England, Vol. 1: *Report,* Cmnd. 4040, HMSO, 1969, pp. 134 and 136.

[8] Royal Commission on Local Government in Scotland, Vol. 1: *Report,* Cmnd. 4150, HMSO, 1969, pp. 246 and 247.

These views were not acted on by later governments. The Royal Commission on Local Government in England had said it believed 'the rate, modernised from time to time, will remain the chief local tax'.[9] Since its yield at the time was second only to income tax, this assertion may not have seemed particularly radical. However, the view was mis-quoted by the then Labour Government and the succeeding Conservative Government; both referred to the Commission's view that 'rates must remain the principal local tax'[10] or 'must remain the principal source of local revenue';[11] and in practice both governments decided that this view justified leaving rates as the only local tax. Shortly after the 1974 General Election, the Labour Government set up a Committee of Enquiry into Local Government Finance, under the chairmanship of Frank Layfield, which reported in 1976. This Committee endorsed the views of earlier bodies. It concluded that the central government

> 'cannot provide local authorities with a preponderant share of their income . . . without sooner or later taking responsibility for their expenditure',

and reported

> 'a strongly held view amongst us that the only way to sustain a vital local democracy is to enlarge the share of local taxation in total local revenue . . . we consider that the administrative cost involved in introducing a local income tax for this purpose would be justified'.[12]

The government of the day, however, was

> 'not convinced that it would be right . . . to add local income tax to existing rates as an additional source of local revenue';

it saw itself as a partner of local government in the provision of local services, and accordingly was prepared to give substantial financial support through grants. It did not think 'any clear advantages would flow from the introduction of a local income tax'.[13]

---

[9] Royal Commission on Local Government in England, *op. cit.*, p. 136.

[10] *Reform of Local Government in England* (White Paper), Cmnd. 4276, HMSO, 1970, p. 22.

[11] *The Future Shape of Local Government Finance* (Green Paper), Cmnd. 4741, HMSO, 1971, p. 4.

[12] Report of the Committee of Enquiry into Local Government Finance, *Local Government Finance,* Cmnd. 6453, HMSO, 1976, pp. 286 and 300-301.

[13] *Local Government Finance* (Green Paper), Cmnd. 6813, HMSO, 1977, pp. 26, 5 and 18.

Whilst there is widespread agreement that central finance is one cause of central control, there is less explanation of why this is so. The main reason is that the central government is responsible to the people who pay taxes to it for the way in which their money is spent. It is inconceivable that any government in 1978 (to take an example) would have been willing to pay over £10 billion in grants to local authorities, a sum roughly equal to a quarter of total central government tax receipts, without exercising considerable control over the way in which the money was spent, and it would seem irresponsible of the government not to exercise control.

It might be argued that the Government in 1978 paid out some £16 billion in transfer payments without feeling it had to control the way recipients spent the money, and that such a policy is acceptable to taxpayers because allowing recipients of transfer payments the right to spend handouts as they choose ensures equal freedom to each taxpayer if and when his or her turn to be a recipient arises. By analogy, taxpayers could confer the right on the central government to pay grants to local authorities without controls, for by allowing other councils to spend as they choose the taxpayer secures equal freedom for his or her own council. However, there are a number of reasons why the analogy with transfer payments is dubious.

First, more than half the money required for transfer payments is produced from national insurance contributions. To a large extent, transfer payments are almost direct transfers from individuals to individuals. If grants to more needy local authorities were largely financed by precepts on less needy ones, there would be much less cause for central control; but this is not how they are financed.

Secondly, grants to local authorities are spent on goods and services which a government will inevitably compare directly with its own expenditure; a government will always be reluctant to hand out money to be spent on schools, minor roads and the police when it is short of money to be spent on universities, trunk roads and defence, unless it fully approves of the way in which the former is spent.

Thirdly, a government no doubt hopes that transfer payments will 'buy' votes for it at the next election, whilst grants to be spent by local authorities are more likely to 'buy' votes for local councillors.

Fourthly, it is in practice easier to control local authorities. If taxpayers see transfer payment recipients spending large sums on alcohol (say), they may feel the taxpayers' money is being ill-spent, but they may accept there is little the government can do. But if they see a local authority spending grant receipts painting lamp-posts (say) at monthly intervals, they may feel the government could and should take action.

Grants and local expenditure

It seems inevitable that governments paying grants to local authorities will seek to control the way in which they are spent. Grants may also curtail the freedom of local authorities to determine their expenditures. The worst possibility, from an authority's point of view, is to have to depend wholly on grants whose value is fixed, an arrangement envisaged for the Scottish and Welsh assembly proposals voted on in the referendums of 1979. In such circumstances an authority has little control over its total revenue (except insofar as it can alter rents or fees and charges).

At the other extreme, we could envisage an authority wholly independent of grants which would find its income (excluding rents, etc.) changing in proportion to its tax rate or poundage. This result would also follow if an authority received a grant which varied in proportion to its tax rate (as does the resources element of the rate support grant). However, the more an authority relies on grants whose amounts do not vary with tax rates, the harder it will find it to alter its expenditure. At present, most grants do not so vary, and it can be said that the lack of local tax revenues and the presence of such grants curtail the freedom of authorities to fix their own expenditures. This in turn makes it hard to change the provision of any service, since, with an inflexible total budget, raising the quantity of one service may be difficult unless another is reduced. Moreover, the situation varies from authority to authority, because some are much more dependent on grants than others.

IV. A THEORY OF FISCAL FEDERALISM

So far, we have implied that grants are paid solely in order to bridge the gap between the income that local authorities can obtain from rates and miscellaneous sources, and the expenditure they (and, presumably, ratepayers and taxpayers) wish to have. We have suggested that the payment of grants will in itself encourage central control, and that the presence of any grants that do not vary with local tax rates will also curtail local freedom. Insofar as grants themselves create a central desire to curtail local autonomy, they seem unsatisfactory; and the correct procedure would be to consider ways of reducing their importance and to ensure that any grants that remain interfere as little as possible with local choice of budget levels.

Three advantages in local government's rôle

There are, however, other reasons why grants may be defended, and we must consider them. Before doing so, we should clarify why we have a tier of local government at all. Why could not all government functions be performed by the central government? In a careful analysis of this question, Professor Wallace Oates[14] has argued that the central government is the appropriate body for

> 'stabilising the economy, for achieving the most equitable distribution of income, and for providing certain public goods that influence significantly the welfare of all members of society',

whilst

> 'subcentral government can supply the public goods and services that are of primary interest only to . . . their respective jurisdictions'.

There are three main advantages in securing such a rôle for a decentralised public sector:

> 'First, it provides a means by which the levels of consumption of some public goods can be tailored to the preferences of subsets of the society. In this way, economic efficiency is enhanced by providing an allocation of resources that is more responsive to the tastes of consumers.
>
> Second, by promoting increased innovation over time and by providing competitive pressures to induce local governments to adopt the most efficient techniques of production, decentralisation may increase both static and dynamic efficiency in the production of public goods.
>
> Third, a system of local government may provide an institutional setting that promotes better public decision-making by compelling a more explicit recognition of the costs of public programs.'

The first of these advantages is enhanced if citizens use the option of voting with their feet,[15] and select for residence the area whose local

[14] Wallace Oates, *Fiscal Federalism*, Harcourt Brace Jovanovich, New York, 1972, pp. 14 and 13.

[15] Charles Tiebout, 'A Pure Theory of Local Expenditure', *Journal of Political Economy*, Vol. 64, October 1956, pp. 416-24.

tax/service mix most appeals to them; also, a system of local authorities may give voters more opportunities to express their feelings on publicly-provided services. The second and third can be seen as combining to reduce X-inefficiency in government.[16]

In such a system, there seems at first sight no justification for grants, for with them will come control, and therefore reduced ability for authorities to tailor services to the preferences of their customers. However, Professor Oates[17] discusses two reasons why grants may be established. First, there may be external benefits from local services, that is, benefits to non-residents, particularly perhaps from services such as education and roads. Local politicians, concerned only with local votes, will ignore such benefits, and so may under-provide these services. In such circumstances, the correct grants are ones specifically related to the services concerned, and they should rise or fall in line with local spending on these services. This argument is persuasive in principle, though one can imagine that valuing external benefits would be hard. However, even if such specific grants were to operate, we would not expect them to be very sizeable in relation to local revenue, and certainly not as much as the present 48 per cent (Table I), since the benefits of local services are doubtless conferred in the main on local residents.

Equalisation grants to ensure 'horizontal equity'

Secondly, equalisation grants are required to ensure horizontal equity. Suppose both central and local authorities treat equals alike; and suppose further that two areas, A and B, seek the same level of local services and have similar efficiency. The rate poundages will be different if the two areas have different rateable values per head: suppose the rateable value per head is higher in A and hence the poundage required is lower. In this case, if Mr Alpha in A and Mr Beta in B have identical incomes, family circumstances, houses and rateable values, then Mr Alpha will pay less in rates than Mr Beta for the same services. A system of equalisation grants to area B could resolve this problem. Moreover, such grants will be required to discourage migration from B to A, and our example contains no reasons to suppose that such migration would be economically

[16] I am indebted to Professor A. T. Peacock for this idea.

[17] *Op. cit.*, Ch. 3.

advantageous, for there is no suggestion that service costs are different in A and B.

In practice, equalisation grants usually take account of variations in 'need' as well as resources. Suppose that areas A and B are as before, except that they now have equal rateable values per head. Suppose, however, that area B needs to spend more money per head than A to provide the same level of service, possibly because it has a higher proportion of schoolchildren and more miles of road per head of population. Once again B will need to set a higher poundage so that Mr Beta will pay more than Mr Alpha for the same services, unless a system of equalisation grants is in operation which takes variations in need into account. Once more, these grants would discourage migration from B to A, and again this is desirable since there is no suggestion that service costs are different in the two areas.

The concept of needs could, however, be extended to cover cost differences. Suppose now that A and B are identical except that A is in a warm part of the country and B in a cold part. In this case B must spend more on heating schools and offices, etc., and again Mr Beta will pay more for identical service levels. The 'needs' part of an equalisation grant could be devised to take this difference into account, but if it did it would discourage migration from high cost B to low cost A. Such factors are not currently considered in determining an area's grant, and there is clearly argument for this present practice in terms of efficiency as well as against in terms of horizontal equity.

Apart from financing externalities and handling equalisation problems, Professor Oates sees no argument for grants. In principle it seems unlikely that the aggregate government expenditure required for these two purposes is high, particularly if equalisation grants were negative for rich/low 'need' areas, for such negative grants or precepts could help finance the positive grants to poor/high 'need' areas so reducing the cost on central government funds. Two other reasons for grants sometimes heard might justify high grants.

Possible virtues of grant finance

First, it is sometimes said that the central government wishes to legislate on minimum standards for local services, since this reflects a widespread desire for uniformity, and grants can be used to ensure such standards since they can be conditional on local authorities meeting them. Moreover, since it is the central government which is requiring the provision, rather than the local electorate, the finance should come from the centre, and local citizens should pay taxes only to cover the extra cost of services above the minimum. This view lay behind the initial financial proposals for Scottish and Welsh assemblies[18] which proposed a system of block grants that could be supplemented by surcharges on local authority rates. It also lay behind some of the discussions of the Committee of Enquiry into Local Government Finance.[19] The Committee was against the idea primarily because of the severe practical problems in defining and policing minimum standards. The theoretical base of the idea is, however, somewhat suspect. If there is really a widespread desire for all authorities to meet minimum standards, doubtless local electorates will ensure they do so without any central intervention at all.

Secondly, it is sometimes argued that grants are a useful way of influencing aggregate local spending for macro-economic stabilisation by demand management. This view has been heard more often in recent years (when grants have been at a record 'high') since macro-economic policies have required restraint on local spending. The view sounds plausible, but there are a number of caveats.

(i) Have countries whose sub-central authorities are less dependent on grants (e.g. the United States) performed less well than the United Kingdom?

(ii) How large a proportion of income has to come in grants for this purpose? The government manages to influence other components of demand such as consumers' expenditure and investment without meeting half the cost.

(iii) It is possible that

> 'it is the level of activity upon capital account which is the largest single long-term influence on revenue requirements; altering the local taxation/grants relationship would not significantly diminish the Government's influence in this respect'.[20]

---

[18] *Our Changing Democracy: Devolution to Scotland and Wales* (White Paper), Cmnd. 6348, HMSO, 1975, pp. 22 and 45.

[19] Cmnd. 6453, *op. cit.*, pp. 286-88.

[20] IMTA Study Group, *Local Income Tax*, Institute of Municipal Treasurers and Accountants, 1969, p. 23.

(iv) High grants may provoke high local spending. The Treasury is quoted by the Royal Commission on Local Government in Scotland as arguing that the

> 'only real solution . . . is for local authorities to raise more of their own finances, so that they have "sufficient incentive to take spending decisions responsibly and to secure economy and value for money".'[21]

If the Treasury advocates lower grants we may infer that there is no need to maintain the present level for control purposes. It is possible, however, that the optimum grant is higher than Professor Oates's arguments suggest, though every increase in the interests of economic control is likely to mean less local autonomy.

V. CUTTING RATE- AND GRANT-BORNE EXPENDITURE

Since local financial autonomy has diminished steadily this century, and there is a case for reducing the dependence on Whitehall grants, is the solution to reduce rate- and grant-borne expenditure? To reduce this expenditure significantly means a change in the provision of a costly service (Table III). Current expenditure on goods and services in 1978 was £17·6 billion, and it fell short of current income by £1·7 billion—the value of the current surplus. Grants were £10·1 billion.[22] In round numbers, current income was some £20 billion of which grants accounted for £10 billion and local sources (rates and miscellaneous income) for £10 billion. The fraction accounted for by grants could be cut from a half to a third (as in the 1950s) either by a cut in expenditure to £15 billion or by levying charges to yield £3·3 billion. The fraction could be cut to a quarter (as in the 1920s) either by a cut in expenditure to £13·3 billion or by levying charges to yield £5 billion.

(i) More efficiency or lower standards?

Expenditure could be cut in a number of ways. Local authorities could attempt, for example, to provide services more efficiently and

[21] Cmnd. 4150, *op. cit.*, p. 246.

[22] This represents a slightly higher percentage than that shown in Table I because Tables I and III employ different sources.

TABLE III

EXPENDITURE BY LOCAL AUTHORITIES IN THE UNITED KINGDOM, 1978

	Current £*m.*	*Capital* £*m.*	*Total* £*m.*
Roads and public lighting	791	329	1,120
Housing	74	1,542	1,616
Refuse collection and disposal	395	38	433
Parks and pleasure grounds, etc.	331	78	409
Town and country planning	166	118	284
Police	1,251	42	1,293
Education	6,344	492	6,836
Personal social services	1,316	60	1,376
School meals and milk	439	—	439
General administration	394	—	394
Other services	1,769	460	2,229
Subsidies	512	—	512
Grants	854	184	1,038
Interest	3,007	—	3,007
Loans (net)	—	–11	–11
TOTAL	17,643	3,332	20,975

Source: National Income and Expenditure, op. cit., pp. 59 and 61.

hence at a lower cost, or they could seek to lower the standards. No-one is likely to oppose a drive for higher efficiency, though it is improbable that an efficiency drive could cut expenditure by the amounts we are seeking. Each of us may have different opinions on the standard of local services, but it is not for economists to tell local electorates what they should demand from their councillors. I do not think that more efficiency and reductions in 'extravagant' services will on their own radically improve local financial autonomy. But it is quite probable that the reverse is true. The Layfield Committee argued that those

> 'who elect local councillors are the best safeguard against waste, extravagance and inefficiency',

and that

> 'they cannot act effectively unless they understand both the rôle of elected members and the financial facts about decisions taken by them'.[23]

Financial understanding cannot be facilitated by a system in which half the expenditure for which local electorates vote is met by tax-payers elsewhere.

(ii) Transfer services to central government

Another method of cutting local expenditure is to transfer some services to the central government. A transfer of schooling would have the largest impact, and Sir John Hicks once mooted the idea of transferring secondary education[24] in the hope that local authorities, facing a reduction in their dependence on grants, would regain responsibility and independence. This seems, at best, a solution of last resort since it would cure the disease only by maiming the patient. It would seem better to ensure that local authorities have enough resources of their own to provide the public goods which (on Professor Oates's model) are best provided locally.

(iii) Raise charges or 'privatise' services

A final way of cutting rate- and grant-borne expenditure is to increase charges for local services. In an extreme case, an entire service (or part of one) might be financed by charges, so that it could, if desired, be transferred to a private supplier. The Layfield Committee considered this possibility and invited the IEA to submit evidence about it; since the Institute has no corporate opinion, the evidence was submitted in a personal capacity by Ralph Harris and Arthur Seldon.[25]

In their evidence they argued the case for charges for at least parts of services such as libraries, fire services, refuse collection and disposal, and personal social services. Speaking about refuse charges, Arthur Seldon summed up the debate neatly as follows:

> 'We are both of us, you and me, conjecturing. We think there are advantages and you see difficulties. I think neither of us can establish

[23] Cmnd. 6453, *op. cit.*, p. 103.

[24] Sir John Hicks, *After the Boom,* Occasional Paper 11, IEA, 1966, pp. 18-19.

[25] Reproduced in *Pricing or Taxing?* Hobart Paper 71, IEA, 1976, pp. 15-54. A fuller discussion of charges is in Arthur Seldon, *Charge,* Temple Smith, 1977.

> his case. I think that is an argument for some local government areas doing it and demonstrating which of us is right.'.[26]

The Committee did not go so far as to favour experiments, but concluded that it

> 'may well be possible for a bigger proportion of local government expenditure to be financed from charges . . . and there could be some benefits from increased efficiency',

and recommended that

> 'there should be a review of both central and local policies towards charging for local services'.[27]

Essentially, this review would try to see if there were any local services sufficiently near the private-good end of the public-good-to-private-good spectrum to make charging the most suitable form of finance. Of course, the more local services that ended up being financed by charges, the stronger the case for a negative income tax to help poor families pay the charges.

Education vouchers only way to cut local dependence on grants

Such a review is certainly to be commended, and any increase in charges that resulted would help reduce local dependence on grants. However, there is no chance of increased charges reducing the importance of grants even to the levels of the 1950s unless they covered education. If they did, parents could meet the cost of school fees with the aid of vouchers which could be financed by the central government. Such an arrangement would vastly reduce rate- and grant-borne expenditure and hence significantly cut the share of grants in local revenue. From the central government's point of view, the cut in grants would be matched by an increase in transfer payments, since we may regard vouchers as a form of transfer payments tied to the purchase of schooling. The central government might still feel entitled to exert some control over schools since it would indirectly provide most of the finance. Its desire for control would, however, be curtailed because the finance was indirect and because parents paying fees would doubtless expect to exert more influence than at present. It is much to be hoped that the present Government, unlike its predecessor, will encourage experimental voucher schemes.

---

[26] Harris and Seldon, *op. cit.*, p. 33.

[27] Cmnd. 6453, *op. cit.*, p. 142.

VI. EXTRA LOCAL TAXES

The only alternative way of reducing grants seems to be extra local taxes. In 1972 I suggested five criteria by which possible new local taxes could be judged:[28]

(1) the yield should be appreciable in relation to current local expenditure;

(2) the rate of tax should be capable of being varied by the local authorities;

(3) the subject of the tax should be widely distributed throughout the country;

(4) it should always be clear which local authority is entitled to the revenue from each individual subject of taxation; and

(5) the tax should not be disproportionately expensive to collect.

Two more criteria noted by the Layfield Committee[29] of Enquiry into Local Government Finance may be added:

(6) 'the level, and most importantly the change in the level, of any local source of revenue should be set locally . . . and the effective burden should not be borne substantially by people outside the area concerned'; and

(7) 'the yield . . . should be able to be varied to match changes in expenditure in line with costs and public policy'.

The size of the gap

The increasing dependence on grants since 1972 means that criterion (1) requires an even more selective approach to possible taxes. Local authorities would have required extra tax revenue of £3·3 billion in 1978 to reduce grant dependence to the level of the 1950s. The only taxes with yields exceeding this were personal income tax (£18·6 billion), corporation tax (£3·5 billion) and value added tax (£5·2 billion); other high revenues accrued from taxes on alcohol (£2·2 billion), tobacco (£2·2 billion) and oil (£2·5 billion before deducting export rebates).[30] No other taxes could make a substantial impact.

---

[28] Alan K. Maynard and David N. King, *Rates or Prices?*, Hobart Paper 54, IEA, 1972, p. 57.

[29] Cmnd. 6453, *op. cit.*, p. 184.

[30] *National Income and Expenditure 1979*, Table 7.1.

The personal income tax does best on criterion (1) and satisfies all the others, though it might cost more to collect than some taxes. Corporation tax and VAT do poorly on criterion (6) and questionably on criterion (4). Alcohol and tobacco taxes raise questions on criterion (2) since people might make special shopping expeditions if tax rates varied appreciably, and this problem would be especially acute if the tax were given to metropolitan districts which are high spenders because they are responsible for education. In 1972 oil taxes or, more specifically, the motor fuel tax seemed promising, but since then they have done badly on criterion (7), for governments have allowed the real yield to fall substantially to cushion motorists against rising crude oil prices.

One tax often considered is a local sales tax, but, as the Layfield Committee pointed out, it would be hard to administer;[31] also, variations in tax rates would have to be small to prevent special tax avoidance shopping expeditions.

There can be little doubt that the local income tax proposed by the Layfield Committee is the most promising in terms of its potential yield, fairly even distribution of yield between areas, administrative feasibility and incidence on local residents.

VII. REFORMS IN THE GRANT SYSTEM

Whether finance for education will cease to be a local responsibility, or whether a local income tax will be introduced, is uncertain. Instead, the present Government is considering the abolition of domestic rates, according to Mr Tom King, Minister for Local Government.[32] For the present, grants are likely to remain important. We saw earlier that local autonomy is helped if grants rise or fall in line with local poundage. At present, only the resources element of the rate support grant works in this way, and this is a small part of the total, as Table IV shows. Its share was increased by the government in 1974-75. Under the provisions of the Bill now before Parliament, the resources and needs elements would be combined to

[31] Cmnd. 6453, *op. cit.*, p. 187.

[32] Reported in *The Times,* 7 May 1980.

TABLE IV

COMPOSITION OF RATE SUPPORT GRANT: SELECTED YEARS, 1967-68 TO 1979-80

	Domestic element %	*Resources element* %	*Needs element* %
1967-68	1·8	16·3	81·9
1970-71	5·3	14·2	80·5
1973-74	6·1	13·6	80·3
1976-77	10·2	29.2	60·6
1979-80	9·5	29·4	61·1

Source: Local Government Trends 1974, CIPFA, 1974, p. 73, and *Local Government Trends 1979,* CIPFA, 1979, p. 24.

form a single large element, all of which would vary with individual poundages.[33] It is possible that the grant would not vary proportionately, and the Minister would have the power to limit severely the extent to which high spending authorities could increase their grants by further poundage rises; but at first sight the new arrangements look promising for those seeking increased financial autonomy for local authorities.

[33] *Local Government Planning and Land* (*No. 2*), Bill 128, *op. cit.*, pp. 33-34.

Arthur Seldon: *That was a good, well-rounded, sober and scholarly survey. We now have a comment by someone who is an economist, a housewife, a farmer and a former councillor, Linda Whetstone.*

COMMENTARY

LINDA WHETSTONE

Dr King has argued that local autonomy is needed if local government is to be responsible and efficient, and also that for local autonomy to be achieved there must be local finance. I agree, but I strongly differ from him on the conclusion of his paper, where the money is to come from, and other points.

How important is it in the national context that local governments become more responsible about money? How many people realise that, during the last five years, local government borrowing doubled the size of the national debt, which now amounts to nearly £1,500 for every man, woman and child, and that the cost of servicing it is more or less equal to the Public Sector Borrowing Requirement (PSBR). A large part of local government expenditure is forced upon it by the requirements of central government, in turn forced on central government by a demanding electorate which does not relate the cost of meeting its demands to the height of its taxes. We must find a way of making those who use the services supplied by local government realise their real cost.

'Irresponsible' grant-aided local government

Grants from central to local government make for irresponsible local government because local politicians do not have to worry where the money is coming from nor account directly to their electorate for raising it. Local taxes would be slightly better than national taxes because the electorate paying the taxes would feel that the levying authorities were more accessible. But local taxes will still not be directly related to the cost of government services to their users. Until we find a system which does so, we will have an increasing demand for such services and ever-increasing local government expenditure and debt. The Wealden District Council, a wealthy Conservative area in Sussex, continues to build council houses because it tells us the housing waiting list has *x* thousand names on it, which indicates there is a serious housing problem. The subsidy to each three-bedroomed house is £26 per week, with rents at about £12. Anyone can understand that when a service is subsidised to this extent there will be a queue for it. But why do not more of the public realise what council housing is costing us in our local areas?

If all council-house dwellers paid a market rent, except in cases of real hardship *via* a means test, and if, better still, at the same time the Rent Acts were abolished, the housing problem in Wealden would vanish overnight. If people were charged the right price there would not be a housing problem tomorrow; people would be moving out of council houses, and no-one would want to move into them. The only reason why they want to live in them is that they are much cheaper than other housing. And as the problem was abolished the pressure on local government to spend, spend, spend would also be abolished.

Council housing is a service where charging would make an enormous difference to spending by local government. Charging for governmentally-provided services is the only way to achieve responsible and economically efficient local government. It is only when people have to put their hands into their own pocket that they seriously consider the alternative uses of the resources. If you put your hand into your own pocket for something, you think about it. If you put it into someone else's pocket, why bother to think about it? It does not matter whether one is a Communist or an arch-Conservative, it is a fact of life. Until our method of ruling ourselves takes the method of paying into account we will be faced with continual expansion of government and continually increasing inflation.

The failure of governments

Sadly, the present Government is not moving in this direction. I am Chairman of the East Grinstead Constituency Conservative Association and I am continually saddened at the lack of progress. The Government has failed to get the idea of charging for school transport accepted by Parliament; the idea was simply to give the *option* to councils—they were not proposing to force councils to charge; and that failed. They have turned down charging for library books; 10p on a library book, for example, would reduce the outgoings of East Sussex County Council by £1 million. Multiply that over the whole country and it would add up to a considerable sum. And what is 10p for a library book? You could charge 20p and it might well yield £2 million, which begins to add up to substantial figures.

The Government turned down charging for nursery schools. When asked by Mid-Sussex Council if they could charge all council housing at an economic rent with a means test for hardship, the Government rejected the proposal without an explanation. Somehow or other we have got to persuade the Government to allow councils to charge because it would make a far bigger difference than Dr King has intimated to education, housing, and many minor services.

Questions and Discussion

JOHN HATHERLEY (*Teacher at King's College School, Wimbledon; Prospective Parliamentary Candidate for the Liberal Party, Carshalton*): In view of the earlier history of the movement in this country, and in view of its success overseas where its adoption is growing, it astonishes me that this country does not examine site-value rating thoroughly. I have heard the most ridiculous theoretical objections to site-value rating which would be completely dispelled if only the critics would go to countries like Australia, New Zealand and elsewhere to see what site-value rating does. There is no inner city decay in their cities, such as Sydney in Australia or Wellington in New Zealand, just as Churchill predicted there would not be. In 1909 Churchill said twice in the House of Commons and in the theatre in Edinburgh that, if we applied site-value rating, the slums of the Gorbals would disappear. The Danes took him at his word and they cleared the slums of Copenhagen in eight years. There would be no inner city decay, domestic rates would be lower and small businesses would flourish. Why will not someone in Britain—and there is plenty of expert knowledge and experience available overseas—investigate site-value rating to see if the objections are borne out in practice?

ARTHUR SELDON: A new source of income, perhaps?

MICHAEL STERN (*Chartered Accountant*): To follow up the point Mrs Whetstone made on the effects of charging for individual services, I agree with Dr King that charges would not raise a great deal of money, but they would have an extremely good side-effect: by forcing people to pay for services, we would find that many would reject some or even most of the services which are currently being shoved down their throats by local authorities. The effect of charging would be to raise a little more money for local authorities, but it would at the same time permit local authorities to cut out a great deal of wasteful, unnecessary expenditure by allowing the public to decide services on an individual basis rather than through a vote every four years or so, which in any case tends to be decided by whether or not the voters like the Prime Minister.

SELDON: Charging will cut costs.

JAMES BOURLET (*Economic Research Council*): On the point raised by Linda Whetstone on council house rents, it would seem that we have to take into account capital gains as well as current transactions. People in the private sector of the housing market receive capital gain as well as getting income tax advantages. It therefore seems to me local authorities have made one of the most sensible investments over the post-war years that they could have made by investing in council houses, certainly better than by investing their money in the Stock Market or in government bonds. I resent the hypocrisy of owner-occupiers constantly trying to cane people who are renting and not receiving the benefits other sectors of the housing market obtain.

ALFRED SHERMAN (*Centre for Policy Studies*): I would like to disagree politely with the last speaker. First, it is not their money that local authorities have put into council housing but ours. They have taken money away from me and you to put into council housing, when they would have done much better to have left it to us.

Secondly, council house are one of the most wasteful forms of expenditure yet devised. Together with rent control, its twin brother, and compulsory purchase orders, council housing has destroyed hundreds of billions of pounds' worth of good houses and replaced them largely by monstrosities. The cost of a council house per unit is roughly five times that of a private house, and the administration costs are correspondingly high, because a privately-owned house has no administration costs. (In Lambeth, for example, it costs £25 to change a washer.

Furthermore, you have layer upon layer of administrators. And all this is not bad merely in financial terms but in resource terms, too, because building is fuel or energy intensive. When you knock down a row of houses, you are knocking down the result of past fuel investment (burning to make the bricks, the transport, etc.). Then you have to spend enormous sums on making cement to make concrete, and again it requires more fuel to transport materials and so on. In the end you merely build bigger and worse slums.

As for capital gain, the owner-occupier has very little in the way of real capital gain. If he moves, he has to spend real money. Most of the increase in the value of his house is purely nominal—paper figures. The local authority, meanwhile, fixes rents according to past expenditure but charges rates on the citizen in terms of current expenditure. The non-council house citizen is therefore being charged extra in order that the council tenant should remain where he is.

One must remember that most council-housing activities are very wasteful, since the beneficiaries are not mainly the tenants—because they, too, as taxpayers and ratepayers, help to subsidise this waste—but the housing committee. I speak as a former councillor involved with housing committees and the housing department.

ARTHUR JONES (*Former councillor; former Conservative MP for Northants South/Daventry, 1972-79*): I have been associated with local government affairs for the whole of the post-war period. There is an increasing necessity to examine the capital commitments of local government. Local government capital assets should be realised, with the sale of council houses as part of that policy. Local authorities should be required to dispose of their vast land and property holdings unless these are required to fulfil their statutory obligations. Substantial savings in borrowing could be made possible by this means, and if local authorities are involved in development schemes—and in some cases it is desirable that they should be—they should not be permitted to hold assets not required to fulfil their obligations.

JOHN FAWKNER (*London Transport*): I am concerned with the planning of capital work on the London Underground. On the subject of providing capital for local authority work, at least in our particular area, the present system of cash limits and the like provides a built-in incentive to what are effectively sub-contractors to local government, like ourselves, to seek government aid even in such areas as the building of office blocks where we can easily raise capital on the outside market.

There is another area where finance, at least for public transport and perhaps other types of services offered by local government, could be attracted in ways other than through rates or from central government. In Paris there is a system whereby employers pay a poll tax according to the number of their employees, which is allocated to public transport; thus employers' decisions in large part determine the volume of public transport which is to be supplied. I do not know whether there are other areas apart from public transport where this could be applied, but it seems that it would have a lot of potential where ordinary individual charging fails to foot the bill.

TYRELL BURGESS (*Reader in the Philosophy of Social Institutions, North East London Polytechnic*): After a study of the delicate relationship between the financial and electoral legitimacy of local government, Tony Travers and I conclude that it is very difficult for local government to be independent without an independent tax base.[1] We also conclude that the activities of the present Government in particular have already seriously threatened to destroy this tax base by the introduction of the block grant. Another colleague of mine, Philip Tunley, has written about the Government's neglect of this tax base through its failure to revalue. The attempts of both Parties to undermine local government by refusing to undertake reasonably regular revaluations is very serious.

What is required is, first, a much more serious examination of local government's tax-raising powers than has been discussed so far today. Certainly there should be regular revaluations which should be required by law—the present Government is abolishing that requirement. Secondly, rateable value should be based upon capital value and not upon what is now an entirely artificial and arbitrary implied rent. Thirdly, we believe the country is grossly under-rated. If the rest of the country were rated in the same way as the South-East, or, at least, if the rest of the country raised rates which imposed no bigger burden, given the ratepayers' incomes, than is imposed in the South-East, there would be no difficulty in financing local government; thus grants could be very much reduced.

It is no coincidence that those local authorities which have been most liable to corruption in various parts of the country are precisely those which are heavily subsidised under the present system and do not raise very much through rates. We would re-rate agriculture, since we can see no ground at all for exempting this prosperous industry from the general level of rates. Furthermore, one could also adopt many of the proposals of the Layfield Committee which have been neglected: local VAT, local petrol duty revenues, and so on.

Finally, we should consider the way grants are distributed. Something that could enormously improve our present position very simply is to get rid of the statistical tomfoolery and general chicanery of the rate support grant by moving at once and very quickly to a distribution of central government grants per head. We worked out in the book in some detail what difference that would make to local authorities' income—and it makes surprisingly little difference. Thus if one simply distributes central government grants per head of population, without having any allocation decisions made by central government at all, and if these grants are much smaller than at present, there is some hope of genuinely democratic local institutions.

---

[1] T. Burgess and H. A. Travers, *Ten Billion Pounds: Whitehall's Take-over of the Town Halls,* Grant McIntyre, 1980.

SELDON: That was a statement of a rather conventional case for increasing taxes wherever you find taxpayers who are not angry. We still have to have evidence that higher tax rates, or higher rates, will yield higher revenue. We must be approaching a point where higher tax rates will not necessarily yield higher revenue.

BARBARA ROBSON (*Islington housewife*): It is very interesting to listen to academics talking about central and local government funding. Unfortunately, from the grass roots, experience tells us differently from the last speaker. In Islington there is a very high rate support grant, probably one of the highest in the country. A borough with only 136,000 people (which is quite small) spends one hundred million pounds a year, most of which comes from central government. However, about £49 million is raised locally. Quite a common rateable level in Islington would be about £500 *per annum*, and many people in comparatively ordinary houses are now having to pay £700 to £1,000 a year in rates.

Local politicians and local council officers can be ruthless in raising rates. Central government does not seem to have much control over the way grants are handled locally. Officers are very adroit at finding their way around controls, and in Islington they are certainly out of control, let alone the expenditure. Every year they borrow somewhere around £50 million, most of which is spent on council housing. The local rates, which are in excess of £40 million, do not even cover the interest charges. Having had rates increases of 32 per cent last year and 42 per cent this year, I am not sure how local government can be controlled by central government. There seems to be no way of doing so at present.

There seems to be an increasing burden on ordinary people, while local politicians and local officers—who are now increasingly politicised, of course—seem to be utterly ruthless in the way they are prepared to use both the local government rate levy and central government grants. It is very nice to take a sort of grand view of the whole country but central London is in a crisis; and to talk about 6 per cent of gross national product accounting for rates and such like in an area like Islington is simply fatuous—in fact very often it accounts for something like 25 per cent of people's incomes. Those are the levels on which most people understand rates and local authority funding.

SELDON: More housewives like that and we'll soon have no rates at all!

P. BUTLER (*Department of the Environment*): An issue that may make the problem even more intractable than has been suggested up till now is that rather more than half of the money raised from local sources—local rates—comes from the occupiers of non-domestic hereditaments. This does throw a rather different light on the general hypothesis that increasing the local tax burden, in contrast to the central tax burden, would increase local accountability.

SELDON: I think our speakers ought now to answer their critics.

LINDA WHETSTONE: I am not so worried about how to raise more money as how to make the best use of the money we have, or how to rely on *less* money. We should not start looking for new taxes—we have really got to find ways of spending less money. Making local politicians more accountable by reducing the proportion of grant would certainly help.

To Mr Bourlet, who disagreed with me over council housing, I would reply that I most certainly do not think that local authorities should be in the business of investing my money. They should be in the business simply of providing houses for those who are at the bottom of the ladder and have nowhere else to live nor means of providing their own housing. Obviously, the Rent Acts have some bearing here, but I cannot agree that we are being unfair to the council-house dweller and that he is not getting the advantage of his investment. It is not his money that is being invested.

Tyrell Burgess seems to be obsessed with raising taxes. We should find more efficient taxes that will reduce the need for more taxes, not be looking for more taxes. We must find taxes which encourage people not to ask the government to spend money—that is the only way out.

DAVID KING: An increasing reliance on charges by local authorities might have a big psychological impact on them. It might lead to reductions in waste, extravagance and inefficiency, but unless we introduce vouchers for schools it is not going to have a significant impact on the contribution by the central government, and until that happens we are not going to get much increase in local government autonomy.

If I interpret the mood correctly, people here seem generally hostile to government, perhaps even to local authorities. I am as willing as anyone to argue that as much should be left to individual decision as possible and be taken out of the jurisdiction of the state, but what is left to the state I want to devolve to the lowest sphere of local authority as much as possible. With competition I can vote with my feet, and therefore I want as little control over local authorities as possible.

Arthur Jones argued that local authorities should be *told* to sell off their land. That may be all very well: a lot of people would like a Conservative government to tell local authorities to sell off their council houses and their land. But how would they react if a Labour government got in and *told* their Conservative authority what to do all the time. My point is that, in so far as local authorities have any functions to fulfil, they should be left to get on with them without central government interference.

Let me conclude with observations on two aspects of local authority taxation. First, the idea of site-value rating has many attractions (one in particular is that it would be possible to get much more sensible valuations), but it also has a particular characteristic which should not go unnoticed: it is a wealth tax. Whereas current local rates, which are based on some notional rental income, are essentially a form of income tax (based on the income you could get if you let the house), site-value rating is essentially based on the value of the house, which is a wealth tax based not on a *flow* of money but on a *stock,* the value of the land. This may be an aspect of site-value rating that people may not like. I am not saying whether or not that is a bad thing, I am merely pointing it out.

Secondly, there seems to be some concern that talk about new local taxes will necessarily mean more taxation. But what people want with a local income tax is to cut the central government income tax since, of course, every £1 a local authority raises from its own tax is £1 less central government has to raise to pay grants. Some of you may think that, in a society where local authorities were wholly dependent on their own resources, they would spend more money than they do now. Mrs Robson, our housewife, seemed to be of the opinion that her local authority was spending much too much money and was effectively saying 'To hell with the ratepayers!'. But people have got votes: why, instead of sitting and bellyaching about it, do not more than 40 per cent of them turn out and vote for a council that would spend less?

PART II
SHOULD LOCAL GOVERNMENT SUPPLY WELFARE? (I)

2. Imperialist Manoeuvres in 'Public' Education and Welfare*

DIGBY C. ANDERSON
Director, Social Affairs Unit

*This paper owes much to the careful and critical comments of Arthur Seldon. I alone am responsible for any errors or infelicities.

The Author

DR DIGBY C. ANDERSON: Education Research Fellow, University of Nottingham, 1977-80. His main areas of research are in the structure of sociological and welfare arguments and the adequacy of the justification for state intervention. His publications include *Evaluating Curricula Proposals* (1980); *The Ignorance of Social Intervention* (1980); *Health Education in Practice* (1979); *Evaluation by Classroom Experience* (1979). He has also written many articles with Dr Wesley Sharrock, published in academic journals, on the media, manoeuvres in arguments and textual analysis. He is currently forming a new research centre, The Social Affairs Unit, to scrutinise the ideas and institutions of state education, health and welfare.

Arthur Seldon: *This is the first of two sessions on welfare services and how they are seen by sociologists. I am grateful to Dr Anderson for helping us to arrange these sessions. Each of these papers will be short because we wanted to find space for two per session. But we shall publish rather longer statements or versions in our* Readings. *Dr Anderson starts this session: he is the Head of a new Unit of which I think we shall hear more, and the editor of a new book.*

I. SCRUTINY OF 'PUBLIC' WELFARE

Wesley Sharrock, June Lait, David Marsland and I are concerned with aspects of the local government and area provision of social, education and health services.[1] I shall refer to all three services as 'welfare'.

'Overgrown' local welfare services

We find many aspects of these services overgrown and incompetent. While not all of us object to all of them, we are profoundly suspicious of them. We see them as novelties which are very much on probation. They are permeated by wishful thinking, inconclusive research, radical ideology, expansionist rhetoric, untested skills and fashion.[2] And thus the suspicions we have of them are not trivial. They would not be quietened by the removal of the Area tier of the NHS, the relegation of the Supplementary Benefits Commission to a local National Insurance Inspectorate or the current and gentle proposals for the school curriculum made by HM Inspectorate.[3]

---

[1] This paper is concerned with how the justification of 'public' welfare projects is managed and not with how often such justifications are used or whether they are used outside the 'public' services. These are proper questions. They are simply not the questions addressed here. Obviously the use of 'welfare' in the paper is very wide and comprehends very different services and practices. The paper does not so much say 'all these services do this' but 'this *may* be *a* useful point to remember when reviewing the case of a welfare service'.

[2] These and other 'ignorances' are investigated in D. C. Anderson (ed.), *The Ignorance of Social Intervention*, Croom Helm, London, 1980.

[3] *A View of the Curriculum*, HMI Series: 'Matters for Discussion', HMSO, London, 1980.

Because many of the aspects of local welfare which arouse our suspicions—generic social work, 'progressive' curricula, health education, polytechnics, etc.—are novelties, the burden of proof in any argument about their worth and right to persist or expand must lie with *them*. They are *social experiments* and it is *their* job to justify their knowledge, qualifications, intrusion into people's lives and their enormous bills to the taxpayers. Correspondingly, it is our job as academic critics, and I expect your's as councillors, MPs, journalists and officials, to make sure that 'public' welfare does evaluate and justify its activities and make available the facts which 'support' their justification.[4] And it is also our job to check their evaluations, scrutinise their facts and audit their claims.

It is at this very early stage that we will fall out with many welfarists. They see themselves as fixtures, subject to a little administrative shuffling perhaps, but, that apart, as the normal, natural, obvious ways of delivering welfare, as permanent parts of national life and budget. Correspondingly *they* see the profound critic as audacious and wild and they require *him* to produce the evidence and claims against them for them to vet and audit. We four critics reject this argumentative sleight-of-hand. *They* are the novelties, the peculiarities and the recipients of people's money. It is normal and right that *they* should have to justify and that *we* should have to vet.

Oliver Twist-ism

And what we have had to vet, until very recently, is a chronic case of 'Oliver Twist-ism'. The welfarists want more of everything: more years at school, more new subjects, more pastoral care, more resources centres, more audio-visual aids, more meetings, more years on teacher training,[5] more in-service training, more social workers, more years on social work training, more liaison, more professional status, more health education officers,[6] more probation

[4] These facts are often difficult to obtain. Their importance in educational debate is discussed in C. Cox and J. Marks, *Education and Freedom*, Kay-Shuttleworth Papers on Education No. 2, National Council for Educational Standards, 1980, p. 2.

[5] The current proposal of the Universities Council for the Education of Teachers (UCET) is *now* talking longingly of extending the Post-Graduate Certificate in Education.

[6] *Health Education: The Recruitment, Training and Development of Health Education Officers*, A Report of a Working Party of the National Staff Committee for Administrative and Clerical Staff, January 1980.

officers, more welfare rights officers, more abortion counsellors, more race relations officials, more new towns, more socially manipulative legislation, more polytechnics,[7] more degrees in leisure, cultural studies,[8] feminism,[9] and marxism, more pamphlets, more 'support' for women, more research, more working parties, more 'relevance', more tenured posts and lots more welfare benefits.

Currently, welfarists' concerns have shifted to protecting what they have got and managing a 'temporary' retreat. Financial stringency has dampened expansionist rhetoric. But the protective and retreating strategies give cause for just as much suspicion as the expansive advances. Though they had considerable warning of the stringency, they have largely failed to evaluate their many 'services' and methodically select those which customers demand ('need'?) the most. The retreat has been as chaotic, unprincipled, hasty, piecemeal and unevidenced as the advance. A wide variation in the style of local retreats has made a mockery of the claims to professionalism and autonomy made particularly by social workers and teachers. The only common and persistent feature has been the effort to save their own jobs. 'No staff redundancies' is the cry. And to protect these staff, services have been reduced, school transport and meals, office opening-hours, part-time staff, etc., cut, lower-level employees such as home-helps or cleaners sacked, institutions, e.g. of teacher training and 'higher education' merged and re-merged, and telephone calls restricted to after one o'clock. Nowhere is there any realisation that the 'cuts' might be permanent. For these welfarists it is the government which is a novelty and an aberration. When it departs, things will, they assume, be restored to their former, 'natural' level. Thus they do not need to undertake a radical and systematic review of their own worth. In welfare, expansion is not a phase but a way of life.

And, while the professionals protect themselves, the bureaucrats 'freeze'. The very jobs which yesterday were claimed to be highly and finely differentiated are today treated as exchangeable in the rush to geriatricise redundancy. Either social work, teaching and higher

---

[7] The columns of *The Times Higher Education Supplement* over the winter 1979-80 raised the possibility of four more polytechnics.

[8] In, for example, the threatened North East London Polytechnic Humanities Faculty.

[9] New Master's course at the University of Kent (1979).

education do have a variety of skills at the professional level, in which case 'freezing' is vandalism; or they do not, in which case . . . There is only one way for a 'caring' profession to cut and that is by evaluating its service to the customer and by sacking its least efficient and successful staff, whether they be Directors of Social Service or cleaners. Such professional evaluation is rarely done, rarely done rigorously and hardly ever publicly available for critical scrutiny.

II. AUDITING THE ACCOUNTS OF PUBLIC WELFARE

When the critic, that is you or I, does inspect the justifications of the welfarists, whether those of advance or retreat, he will wish to audit them both economically for cost, opportunity cost, etc., and socially to see whether they really do show a social products. Because welfarists are not accountable to the customer through the market it is essential that there are critics to demand and audit the accounts of 'public' welfare. The critic will obviously want to study the facts: relative examination pass rates of schools, social work case-loads, the ratio of bureaucratic to customer-contact hours, and so on. These are, as it were, the entries of the account. But the critic should also inspect the way that the welfarist uses these entries to arrive at his result, his justification. We wish to draw your attention to 11 of these manoeuvres with 'facts';[10] and we take them not only from the justificatory documents of welfare but from the verbal arguments of teachers, social workers, lecturers, health education officers, etc.[11] I shall quickly list them: my colleagues will elaborate on some of them.

1. Normalisation

The first manoeuvre we have seen already. Let us call it *normalisation.* The welfarist seeks to evade scrutiny by representing his request

[10] Arthur Seldon talks of some matters which are very akin to these manoeuvres —'unfinished business' and 'gaps'—in his recent rigorous exposé of the NHS: 'The Next Thirty Years?', Ch. 7 in C. M. Lindsay *et al.*, *National Health Issues: The British Experience*, Roche Laboratories, 1980, pp. 108-9.

[11] These justifications function not only to convince outsiders and officials through documents but, expressed casually and verbally, sustain members' self-confidence and are part of the mundane daily talk of such members with each other and their clients, wives, etc.

as normal. The more years this is done the more the item moves from being contentious and new to being non-controversial, expected, established. It becomes an institution, closely enmeshed with others, and acretes all sorts of vested interests. Finally its rejection becomes unthinkable. The critic should remember that this 'normal' and 'established' practice, whether mixed-ability classes or social work in-take teams is perhaps no more than four years old, was possibly introduced as an experiment and that repetition is not the same as success. There is a need to stock-take not only the new experiments of last year but the repeated experiments of the 'sixties.[12]

2. Closing

Often the justificatory evidence the critic will be given is social research. Such research should be treated with caution. It rarely implies any *one* policy. If it does so it usually suggests what to aim at rather than how to get there. In a recent study of Schools Council curriculum projects I found their claims to be 'the result of research' at best metaphorical, at worst pretentious.[13] The manoeuvre consists in *closing* the gap between the policy and the research, making it appear that the latter justifies the former. In practice that gap is often plugged by ideology and wishful thinking.

3. Spreading

Even if research justifies a policy, there is always the question of how big a policy it justifies. One of those Schools Council projects,[14] for example, makes out a good case for health education in schools. What it does *not* do is to justify its proposals for 11 years of health education. Similarly, there may be a disproportion of *time*. A recent report on health education in the NHS[15] confidently and uncritically presumes an increase in the number of health education officers from 234 (1976) to 810 (1988), expecting 550 in 1980.

[12] The problem of fossilised novelty applies to ideas as well as to organisations, especially academic ideas taught on applied courses. A discussion can be found in A. G. Thomas, 'Education and Ignorance', in D. C. Anderson (ed.), *The Ignorance of Social Intervention, op. cit.*

[13] D. C. Anderson, *Evaluating Curriculum Proposals: A Critical Guide*, Croom Helm, London, 1980.

[14] *Schools Council Health Education Project 5-13,* Nelson, London, 1977.

[15] *Health Education . . ., op. cit.*

What is happening with health education both in schools and the health service, or rather what these people would have happen, amounts to *premature institutionalisation.*[16] The desire to launch into massive schemes on puny evidence and to establish such schemes everywhere is endemic in education and welfare. Why were health evaluation officers not called 'Experimental Health Education Officers', appointed for five years in six areas only and monitored?

4. Obligation

Enormous schemes are often justified by alleging that there is an enormous 'need'. Though speaking for someone else's 'need' is always a hazardous business, pleas for colossal social engineering such as the Royal Commission on the NHS[17] or the Seebohm report[18] have no qualms about grounding wholesale social change in attributed 'need'. The critic will remember that even if one could be sure of 'need' there is little point in widespread action unless there is some reasonable *solution.* Despite much wishful thinking, there are still problems we do not know how to solve. 'Need' is not a *sufficient* justification for public expenditure. Where are the tested skills to match the awful 'needs' which 'justify' the Manpower Services Commission[19] or the Commission for Racial Equality or the social engineering of the 'new cities'?

5. Theorising

Welfare is not done to someone. It involves, indeed depends for its success on, the response of the customer. But in some welfare plans the customer does not figure at all except as some sort of dope[20] who is assumed to act in a theoretically 'logical' way. Rent Acts, housing improvement grants, planning blights, inner city areas and the

[16] Discussed in detail in 'Some Preferences' in D. C. Anderson and E. R. Perkins (eds.), *Practical Prospects for Health Education in the Eighties,* a report of the Leverhulme Health Education Project, University of Nottingham, 1980.

[17] *Royal Commission on the National Health Service,* Cmnd. 7615, HMSO, 1979.

[18] *Report of the Committee on Local Authority and Allied Personal Services,* Cmnd. 3703, HMSO, 1968.

[19] A. G. Thomas, 'Education and Ignorance', in D. C. Anderson (ed.), *The Ignorance . . ., op. cit.*

[20] H. Garfinkel, *Studies in Ethnomethodology,* Prentice Hall, Englewood Cliffs, N.J., 1967.

dislike of employers for the CSE are superb testimonials to the abilities of customers to be more various and human than the models of planners would allow. The manoeuvre of theorising consists in passing as complete a plan which does not allow for customer reaction except in a model way.

6. Evaluation

Evaluation is a powerful weapon in the argumentative armoury of the welfarist if he is allowed to use it when it suits him. First he may use it to turn even failure into a friend: 'The problems with our work with juvenile delinquents arose not because we were given too much but too little power', some social workers will argue. The failure of this or that scheme can always be turned into an argument for more money, power, coverage or whatever. If one is to obtain a better evaluation than this it is essential that the critic requires the welfarist to state *before* action what the latter will accept as indicating incompetence, waste and failure. While it may be difficult to evaluate 'work with people' in a precise way, it is reasonable to ask the planners of Milton Keynes or the advocates of comprehensives what they would take as evidence of failure, how they would set a standard of *minimal* attainment.

Manoeuvres 7-11

Selectivity is also to be watched in appeals to moral obligation and compassion (7). It occurs in the claiming of professionalism by the NUT or BASW in order to demand exclusive rights and their inconsistent unwillingness to evaluate and police their colleagues' work (8).

Another manoeuvre has to do with vocabulary. The names, skills, posts, organisations and qualifications are designed to display their owners as 'caring' and 'specialised'. The critic who, using their language, attacks the owners, finds himself, like the government, cutting 'essential services' to 'children in need', depriving the 'poor' of the 'skilled care' of 'qualified' 'social workers', vandalising the 'dedicated' work of 'experienced teachers', etc., etc. (9). Another example is in the use of 'community', especially in 'community health' (10).[21]

[21] Explored in T. Packwood, 'Community Care: The Universal Panacea', in D. C. Anderson, *The Ignorance . . ., op. cit.*

I have not forgotten the manoeuvre which puts the evidence for action beyond the critic's reach in some foreign discipline,[22] nor that which splits up its pleas into acceptable isolated units, arguing piecemeal for that which would never be countenanced whole. There are those who would use welfare for wider political ends (11).[23] In summary, the scrutiny of 'public' welfare proposals will not be attained by a gathering of facts or rival facts but depends on close attention to the way facts are used and manoeuvred in documents and discussion.[24]

---

[22] D. C. Anderson, 'Borrowing Other People's Facts: The Case of Social Inquiry Reports', paper read at the British Sociological Association's Annual Conference, University of Warwick, 1979.

[23] Classically documented in higher education by J. Gould, *The Attack on Higher Education: Marxist and Radical Penetration*, Institute for the Study of Conflict, London, 1977. A more recent display of such influence is Macmillan's current list in sociology.

[24] A more detailed guide to argumentative manoeuvres in the social sciences is in D. C. Anderson, 'Practical Issues in Writing and Winning Sociological Arguments', British Sociological Association's Annual Conference, University of Lancaster, 1980.

3. Some Problems with Social Research

W. W. SHARROCK

University of Manchester

The Author

WESLEY W. SHARROCK: Senior Lecturer in Sociology, University of Manchester, where he has been since 1965. He has contributed to books and journals, including R. Turner (ed.), *Ethnomethodology* (Penguin Books, 1974); J. Schenkein (ed.), *Studies in the Organisation of Conversational Interaction* (Academic Press, 1978).

Anyone looking to social research for assistance in policy-making will wisely regard with suspicion any political conclusions which may be offered as the implications of the research. Much social research does not imply any specific policy conclusions and does not favour one side of a political argument over the other. Where political conclusions do seem to follow directly from the research it is sensible to check that they are not functioning as both premisses and conclusions of the same argument.

The idea of a policy science no doubt seems like a good one, enabling the decision-maker to place decision-making on a more rational footing, making decisions easier to take and more effective. The consumers of social research—administrators, policy-makers, etc.—do not want to be troubled by the theoretical, methodological and technical problems of the disciplines upon which such research draws, but they remain in ignorance of these 'technical' matters at their peril. They ought, at least, to be aware that problems do exist, and that they are serious. I will mention four problems in sociology.

(i) **Problems with measurement**

Those of us who engage in what is often disparagingly called qualitative research are often enjoined to adopt quantitative methods, to abandon vaguely quantitative expressions like 'some', 'many', 'a few', and to employ properly scientific measurement which will produce precise findings, exact expressions of quantity. Quantitative research does have the superficial appearance of more genuinely scientific work, and to those outside the discipline who are paying for the research the presentation of large quantities of statistical tables and acres of computer print-out may appear to be evidence that their money has been invested in something substantial. They may well be unaware that there are often severe difficulties in making genuine measurements.

There are two main kinds of problem with measurement, those that have to do with the 'operationalisation' of concepts, and those having to do with the measurement operations themselves.

The problems in operationalising concepts are those that arise from devising procedures that will correctly identify and measure the quantities we seek to describe.[1] Continuing contention over the testing of intelligence exemplifies these difficulties. Although there are scales which purport to measure intelligence, there are serious doubts as to whether they do indeed assess activities which can be taken to display comparative intellectual abilities.[2] Even if one can devise agreed scales, there are often chronic problems in collecting a sufficiently large, representative or cogent quantity of data. Thus a recent, large, and quite costly study of social mobility in Britain was based on a large sample, but one composed of males when a realistic general assessment of mobility rates would necessarily include females.[3]

Even if the figures which research collects are sound, they are not usually all that precise. They are statistical in nature, in most cases, and are therefore governed by rather large tolerances for error. If, say, a figure of 52 per cent is given for the value of a variable, it means only that the figure we are talking about could be anywhere between (say) 48 and 57 per cent (if not in some wider range of variation). It is often not possible, however, to accept that the figures are sound, both for the reasons mentioned above and because they may represent the result of attempts to quantify the incalculable, consist of guesswork and extrapolation rather than measurement. Some sociologists (and some policemen) have known for a longish time about the inadequacies of attempts to make estimations of crime rates on the basis of police statistics, since such statistics tell at least as much about the information-collecting activities of the police as they do of the nature of the activities they record. The point can be generalised to many kinds of official statistics.[4]

(ii) **Problems in interpreting the research**

Even if the measurements researchers made were precise, we should not necessarily be a great deal better off, for information is only as

---

[1] A comprehensive discussion of sociology's methodological problems is in A. V. Cicourel, *Method and Measurement in Sociology*, Free Press, New York, 1964.

[2] For example, P. Squibb, 'The concept of intelligence', *Sociological Review*, Vol. 23, 1971.

[3] J. Goldthore, *Social Mobility and Class Structure in Modern Britain*, Oxford, 1980, and A. H. Halsey *et al.*, *Origins and Destinations*, Oxford, 1980.

[4] A. V. Cicourel and J. Kitsuse, 'A note on the uses of official statistics', *Social Problems*, Vol. 11, 1963-4.

good as the questions that are put to it. Social analysis is not, at present, a very refined instrument, sensitive to slight variations in quantity. It does not, on the whole, matter much whether the value of a variable is, say, 47 or 52 or 58 per cent, since it will be interpreted as, say, less than expected, more than chance would indicate, around a half, or in some equivalently crude way.

Assessment of the significance of such figures depends upon the standards that are brought to bear upon them: how many is 'a lot', how much is 'too much'? The kinds of questions currently put to data are often the bluntest kind, inviting choice between alternatives of the grossest variety, which means that extensive, detailed and precise figures are not really necessary to a choice between them. The mobility research mentioned above asks whether education can change society, and presents the alternatives between which choice is to be made as being those which see education either 'as a rock on which a modern and prosperous civilisation is built'[5] or as 'essentially an organisation of control by one generation over the next'.[6]

(iii) Problems in identifying the problem

One respect in which it is necessary to be very careful in interpreting social research is in identifying the problems to which it is addressed. Researchers are prone to re-conceptualise the problem. To obtain the required operationalisation, they often feel it necessary to define crucial concepts to make them (what they regard as) more precise or objective. Terms which have a widespread common usage are re-defined, which means that their purported sociological use will no longer coincide with the common usage. Thus, a recent study of 'poverty' begins with the claim that

> 'poverty can be defined objectively and applied consistently only in terms of the concept of relative deprivation'.[7]

To define it in that way is to sever the connection between its new technical use and the conventional understandings people have of the term. It would seem sensible, if such re-definition *is* essential, that the term 'poverty' be dropped and the expression 'relative deprivation' substituted. That, however, would deprive the author

---

[5] Halsey *et al., op. cit.,* p. 2.

[6] *Ibid.,* p. 3.

[7] P. Townsend, *Poverty in the United Kingdom,* Penguin Books, 1979, p. 31.

of the report of the opportunity to make the claim that poverty in the United Kingdom is more extensive than is generally or officially believed. The conventional sense and implications of the term 'poverty' are traded upon to carry implications that the technical term would not.

(iv) **Problems with time-scales**

At present, sociology is prone to rapid changes in opinion. Enthusiasm which is unqualifiedly endowed upon one position will very soon, and equally wholeheartedly, be given to another. Who now remembers the 'new sociology of education' which, basing itself upon some slight studies of life in classrooms and an equally slight study of small boat navigation in the Pacific, made vast claims for its own significance, rapidly published a large number of boasting and self-congratulatory papers—and disappeared from public view?[8]

There is an important difference in time-scales between researchers and policy-makers. For researchers, rapid and frequent changes of outlook are not terribly serious, for they chiefly involve giving up a theory, and that may be no sacrifice at all. The theory's fortunes may be declining, and as often as not the researcher will have done little to work the theory up, probably simply having taken up somebody else's ideas. Changing theories can count as a virtue, for it can indicate open-mindedness and flexibility, and thus be counted as intellectual progress. Changes of policy direction do not take place quite so quickly and have consequential aftermaths. People can find themselves living with the sad results of once-fashionable theories which may have been abandoned in the academy almost before they entered policy thinking.

The possibility of radical changes in direction is enhanced by the relative vagueness of a subject like sociology. Its ideas are broad, loose and, usually, hardly thought out in any thorough, consistent or careful way. This means that quite radically opposed policy implications can be seen as deriving from the same set of premisses, one interpretation prevailing for a time, the swing between them being something of a pendulum movement, taking place when it begins to seem that that interpretation has been overdone and is in need of correction. We are perhaps beginning to see something like

[8] For example, M. Young and G. Whitty, *Society, State and Schooling*, Falmer Press, 1977.

this movement in the case of the family which, for a time, was the villain of the piece, obviously oppressive, destructive of human identity, driving people to madness, essentially an instrument for disciplining people to the requirements of capitalism.[9] Now matters seem to be going the other way, and it is the meddling interference of state bureaucrats with the autonomy of the family which is the evil, obviously done in the interests of disciplining people to the requirements of capitalism.[10] That one can as well argue either approach on much the same premisses and evidence might seem grounds for diminishing the ardour with which either is pressed, but this is not the conclusion usually drawn.

Broad, longer-term swings in sociological fashion

The swings which go on within contemporary sociology are perhaps part of a broader movement over a longer time-span. Not long ago, in the post-war period through to the early 1960s, it was fashionable to look for 'good news', to find that things which were ostensibly useless (e.g. religion) or bad (e.g. organised crime) were either good or useful or both.[11] Now the fashion is to look for 'bad news', to find that things which appear good, harmless or utterly inconsequential are in fact bad and damaging and serving an oppressive function.[12]

The ways in which the benefits and drawbacks of social practices are established will not, I think, usually withstand close inspection. One might think that if people were seriously interested in assessing the value and rôle of institutions they would seek to devise ways of auditing the credits and the debits. But researchers tend currently to be single minded, looking for virtues or faults alone. In consequence, contradictory pictures are painted. In the 'fifties and early 'sixties, an unbelievably rosy picture was painted,[13] its excesses perhaps encouraging the reaction which now leads to a portrait of unrelieved

[9] A review of some relevant issues is in Mark Poster, *Critical Theory of the Family*, Pluto, 1978.

[10] For example, Christopher Lasch, *The Culture of Narcissism,* Warner, 1979.

[11] For example, R. K. Merton, *Social Theory and Social Structure,* Free Press, 1957.

[12] For example, H. Marcuse, *One Dimensional Man*, Routledge, 1964.

[13] For example, D. Bell, *The End of Ideology*, Free Press, 1960.

gloom and oppressiveness. In assessing sociological accounts it is necessary to bear in mind they do not usually tell very well-rounded stories.

The current tendency to treat sociology as though it were an extension of left-wing ideology will not be repaired by another swing of the pendulum, the displacement of one set of political prejudices by another. One of the founders of modern sociology, Max Weber, argued for the separation of the political and scientific aspects of social thought.[14] His arguments are not to be dismissed on the trivial grounds that facts and values are not easily distinguishable, for Weber's argument was founded in the recognition that they were not. It has recently been easy in sociology to argue that sociology is necessarily political, but the popularity of that view does not make it right.

Questions and Discussion

ALFRED SHERMAN (*CPS*)**:** Whilst I agree with all that Digby Anderson said, I think he missed out one factor—that the word 'welfare' is a weasel word. It is a lie word. The original meaning of 'welfare' was 'well being', 'faring well', whereas in its current meaning it means 'giving private goods and services free or subsidised at the point of consumption'. I say 'subsidised', but it does not necessarily mean 'cheaper': a council house, though heavily subsidised, can be very expensive because it is so wasteful. All those goods and services which are handled by so-called 'welfare'—education, health, housing, etc.—are private goods for which other people pay. If you send your son to Eton, after all, you are not considered to be buying welfare. Some of the goods and services consumed by individuals may benefit the population, but this is also true of soap because if people do not wash . . . But we do not regard the provision of soap, or even food, as welfare.

SELDON: Any more views on how far welfare is a public good and therefore ought to be supplied by government, and how far it is a private good and ought not to be supplied by government, central or local? I think that is an issue worthy of a number of comments.

---

[14] Max Weber, 'Politics as a vocation' and 'Science as a vocation', in H. H. Gerth and C. W. Mills (eds.), *From Max Weber,* Routledge, 1948.

JOSEPH EGERTON (*Association of British Chambers of Commerce*): One topic which deserves some attention is the extent to which local authorities are beginning to set themselves up as industrial or employment promotion agencies. The Inner Urban Areas Act facilitated originally and now there are new proposals for ever greater powers to spend ever larger amounts of money allegedly on bolstering up local industry and commerce. Quite often, of course, the reason why local industry and commerce is in a bad state is either that the local council has been inept in its planning policies and has so over-regulated land use and building policy in its area that no-one wants to set up there, or alternatively it has been so exorbitant in its rate demands that people have either moved to another area or simply gone bankrupt or shut down. In response to this, the local authorities, instead of reducing the rate burden and making the area more attractive, have set up ever larger bureaus of officials who think they know what small firms need—indeed, on occasion they think they know what big firms need.

This is the latest growth area in a long list of what is allegedly called 'welfare' and one that is so obviously and so utterly wrong and at which local authorities are so singularly inept and incompetent that the Government should be told firmly that not merely should there be no further extensions—and extensions are being considered—but that there is a very good case for pulling the local authorities right back out of this area, putting a barrier round it and saying 'Thus far and no further'. This is, after all, one of the areas where we are seeing the duplication of effort between central and local government which is inherently wasteful.

PROF. JULIUS GOULD (*Department of Sociology, University of Nottingham*): I know that Dr Anderson is interested in techniques of study and, unlike most of the sociological tribe, is conscious of the difficulty of formulating questions and the caution required in producing answers. A topic that is germane to this Seminar is what was called the 'politicisation' of local authority officials. This is a very important and recent development connected in some ways with the union structure and the growth of white-collar public service unions. What I would like to put on to Dr Anderson's agenda is the need to produce a usable manual for the study of 'politicisation', and a set of questions, and perhaps also attention to the techniques that would be appropriate in answering those questions.

The importance of this development is underlined by what many people here may not have come across: the great attention that is being given—and has been given for four to five years now—by the various groupings of Marxists in this country, and especially the strongly-revived Communist Party, to what they call the 'local state'. Great attention is given by the more coherent, better organised and non-anarchist Left to the understanding of what they call the 'state' and to assessing its significance in relation to the national state with which, of course, Marxists and others have been long pre-occupied. There is a need to study the 'local state', to use their terminology, but to do so by appropriate techniques and questions. One set of questions, I suggest, relates to 'politicisation' and to its impact on the provision of services.

P. BUTLER (*Department of the Environment*): I would like to raise what I regard as the dilemma that has emerged here. A previous speaker has advanced arguments for setting boundaries around the activities of local authorities because they are regarded as wasteful. Earlier speakers argued that there was a good deal to be said for devolving responsibility to local authorities to make them answerable to the local electorate. Aside from the question of who pays and the extent to which the local community should pay for its decisions, I think there should be more debate about the issue. The only body likely to start putting boundaries round local authorities is central government: where does that leave our centralisation/decentralisation model?

DAVID MARSLAND (*Department of Sociology, Brunel University*): That is presented as a difficult dilemma, but if you decentralise from the central to the local state you can kick out of the local state back to the citizens things that do not need to be done by any state at all.

PROF. NATHANIEL LICHFIELD (*Emeritus Professor of the Economics of Environmental Planning, University of London*): In environmental planning we are concerned with doing many of the things that Dr Anderson criticised. I should like to respond to his eleven-point attack on the welfarists, which seemed to emphasise that somehow the welfare approach, whether in inner cities or new towns, arose from people's needs. I would suggest that, certainly in town planning, these programmes—if you can call them that—come out of what are recognised as problems.

After the war it was seen that people were going to need new homes, whatever their income or social class. It was decided that one way of meeting the demand for new homes was to build new towns rather than add them to the edges of cities. Though we are still not sure whether that was a good solution or not, the housing was not seen as a 'need' to be satisfied in the welfare sense. It was seen as a problem, and the planners thought that if nobody had taken action on whether to build new towns or houses on the edge of the city, there would have been a shortage of homes. That was the approach.

Now I want to make a comment about the inner city areas as the latest 'manifestation of this ridiculous exercise'. It is argued that the inner area problems of local industry are due to the planning controls, etc., of local government, and that if local government would get out of the act everything would be fine. The present Local Government Bill has gone some way towards this solution with what the Chancellor calls 'Free Enterprise Zones'. From my two-year study at the Department of the Environment on what we call 'local economic development planning' (and this is simply one piece of research evidence among many), what is really happening to employers in industry has very little to do with local government, although one can find examples where the local authorities should be less strict. The problem basically goes back a long time before local government planning powers were at all effective, to the origins of those cities, the era in which they were built, the structural problems of British industry. These problems have found geographical expression in the inner city. And that is the starting point.

If it were thought that the market could cope with the problem, then the market solution might be tried. But the conclusion of the studies that have been going on for a long time is that something over and above market forces is needed. You can, of course, criticise whether the application of the planners' controls has been good or bad. But it was necessary.

Evaluating the impact of welfare schemes

I come to the evaluation points that Dr Anderson made. His requirement is for those who put up these welfarist and other schemes to justify and evaluate their impact. But that is part-and-parcel of practice in this area. We do try to evaluate the impacts among alternatives and reach a conclusion. Then you reach a stage where Dr Anderson's comments are helpful in studying what those impacts have been in practice, because no-one who tries to do complicated predictions about the future can be at all sure that he is on the right lines: the only way to find out is from research control or monitoring of the kinds Dr Anderson talked about—of what goes on in practice as a basis of learning.

As a planner I would accept the statement that the market could do more than it is doing today, and, in addition, I would accept the statement that bureaucracy is doing more today than it should do. But we are still left with a vast amount of things which have to be done by some level or other of state enterprise. Even if you give a great deal of freedom to what is called 'community enterprise' (people on the shop floor), which I am all in favour of, you are still left with many hard chestnuts like the new towns and the inner city problems which are not going to be solved through market forces.

So the question is: What do you do? Here I find the approach of most economists negative, for while conventional economics has built up a large body of knowledge on prices and markets in what is called 'private sector economics' over the past two centuries or so, there remains a considerable need for the development of what is called 'public sector economics' to assist public authorities locked into 'situations of inevitability' to get better value for money for what they have to do.

Everyone here would surely support the need for the more efficient use of the resources spent by the public sector if they are in any event going to be spent. Therefore, there is a need for the recognition of the rôle of government in this matter. I do not believe Professors Friedman and Hayek—or even Adam Smith himself—would disagree with that statement. What must be done is that we should try to help government to perform its rôle better.

Dr Anderson has made some penetrating points from a sociological standpoint, but if they are simply sticks with which to beat the welfarist, then they are, of course, interesting and I have learned from them because they are valuable tools. But that is not good enough. You must find positive measures by which to help the welfarist (your term) to do the positive jobs that have to be done.

FRANCIS WHETSTONE: One thing one must always bear in mind with the provision of welfare by local government is its general debauching effect: the more the state provides, the more people look to the state or local government to provide. One finds that particularly with people who could be expected to be able to look after themselves. The people who complain the loudest about the cuts are those who are doing evening classes in some subject which they could well pay for, or people who like the local theatre subsidised. Continuous welfare feeds on itself.

DIGBY ANDERSON: There is perhaps evidence of a small misunderstanding here that ought to be put right. Questions about the local provision of welfare and education should be separated into economic and social ones, and I argued briefly, first, that it was important to ask social questions about social matters. One should consider whether things which claim to have a social worth do have a social worth, as well as how they should be financed. My complaint was not that these services did not deliver the goods, but that we do not *know* whether they deliver the goods. Their administration does not help us to find out whether they deliver the goods effectively. In particular, on the issue of planning, it may very well be that the planners themselves evaluate their activities with extraordinary finesse, but as far as allowing the public information and scrutiny of predicted standards, so that one can see if people are failing, the debate is a very uneven one at the moment. It is one in which the critics enter with an enormous disadvantage, and that is what I was complaining about, not the actual provision of welfare. We are not yet in a position to say just how good or bad it is.

W. W. SHARROCK: I am not saying that the sociologists and others I argue against are wrong and I am right. What I would maintain is that all sociological positions are basically arguable and that there are some pros and some cons for most of them, including my own. What has been a chronic problem for me in my discipline is that many of my colleagues simply foreclose discussion by saying: 'Something needs to be done'. The need to 'do something' is simply used to encourage the adoption of whatever policy they happen to favour at the time.

I would like to see much more caution in the attempt to move from academic ideas and research to the drawing of policy implications of any kind. That is not to say that there are not things that need to be done and that they should not be done. The delusion is to suppose that in practice the use of research will improve the decision-making. I suggest that often it may not. It may be better to say: 'Well, we have to do something; we don't know what the right solution is but we have to do something', than to say: 'Well, we have to do something and here is research that tells us the right way to tackle the problem'. One may then follow that research and find that, after all, its implications and conclusions were incorrect. So I am simply arguing for much more caution in the move from the academic base of social science to the drawing of policy conclusions of any kind.

PART III
SHOULD LOCAL GOVERNMENT SUPPLY WELFARE? (2)

4. Central Government's Ineptitude in Monitoring Local Welfare

JUNE LAIT

University College of Swansea

The Author

JUNE LAIT: Lecturer in Social Administration, University College of Swansea, since 1970. Author, with Dr Colin Brewer, of *Can Social Work Survive?* (1980). She contributes regularly to *The Spectator, World Medicine, Community Care* and the *Daily Telegraph*.

I. WHO WANTS/'NEEDS' 'SOCIAL WORK' SERVICES?

Over the past 20 years, and possibly longer, local authorities have been saddled with functions that are impossible, and since they are impossible, can hardly be said to be necessary. Necessary is a word one should use with great caution, possibly never in the context of local government welfare services. The services I wish to discuss are provided under the Social Services Act, 1970, and in particular those going under the label 'social work'.

Doubtless public opinion has changed since the 1840s when *The Times* remarked that citizens preferred to take their chance with cholera rather than submit to public regulation of water supply and sewerage. It may not have moved nearly so far as involved professionals, trained to provide a service and understandably anxious to undertake work 'commensurate with their training' (as they would put it), would like us to think.

There is an uncharted—and possibly unchartable—relationship between provision of services and people's desire to use them, but the case study I wish to present, social work in the London Borough of Tower Hamlets, does not indicate that the supply of 245 social workers caused the citizens of that borough to require their services, for when these functionaries (members of the Department of Applied Love, as the *Daily Telegraph* once called them) were on strike for 10 months recently, no-one seems to have noticed.

Duty of inspection an important safeguard

When I first taught a university course entitled 'The Development of Social Policy 1834-1938' I used confidently to assert (along with the major texts on the subject, and such direct evidence as I was able to assemble) that the Revised Code of 1862, which placed on Inspectors of Schools a duty to inspect registers and ensure that children were proficient in the 'Three Rs', served to sour relations between teachers and inspectorate. I burbled on about restricting the talents of inspired teachers with mundane matters like there being as many pupils actually present as there were ticks on the register, and the iniquity of basing pay on crude measures like the numbers who could read and write. Now, as with so many of the comfortable liberal, not to say 'wet', convictions of my youth, I am not so sure. I begin to see the teachers' and more trendy inspectors' protests against routine, possibly boring, but undoubtedly necessary functions as possibly an early manifestation of the capacity of 'welfare' professionals to redefine their jobs in terms of what they enjoy doing rather than submit to measures of what the customers or the paymasters have asked them to do. Victorian teachers were not nearly so adept at obfuscating jargon as are modern social workers, nor had they staked a corner in compassion for themselves. They had to sell their wares to a public suspicious of state intervention, wanting value for money. One of the functions of Her Majesty's Inspectorate was to see they got it.

How dramatically we have progressed since those benighted times! Or even, to emerge from the mists of history to the working lifetime of the present elderly speaker, in 30 years. As long ago as that, inexperienced but endowed with the then novel qualification of 'Child Care Officer' (acquired after a year's academic study of hilarious irrelevance), I was judged capable of attending to the 'Child Care' needs of a Midlands borough of 100,000. I was supported in this venture by an elderly lady of great benevolence but no qualification save the university of life. Between us we carved up the city and set off on our bikes to do good to those of the city's children whose needs were served by the Children Act 1948. We were charged with visiting children fostered in private homes at least every six weeks. We did so, we recorded our visits in longhand, and fitted in a variety of other activities such as appearing in court, speaking to women's organisations about our work, arranging adoptions, finding foster homes, and occasionally negotiating with housing departments and the National Assistance Board (as it was then called—now the Supplementary Benefits Commission). Withall we did not feel especially underpaid or overworked, and when the Child Care Inspector came unannounced to see us he found our records, which he checked carefully, up to date. What he thought of the quality of our work he did not disclose, at least to us, but we stayed in post until we wished to leave, so perhaps the inspector did not expect either of us to solve all the problems of unhappiness, idleness or downright greed in the respective halves of the borough for which we were responsible, as well as doing the job for which we were appointed.

II. THE TOWER HAMLETS SOCIAL WORKERS' STRIKE

All 245 field social workers, repeat *245,* not counting Seniors, Principal this, that and the other, Directors and Assistant Directors, Area Whatnots, etc., etc., who served the needs of the 150,000 population of Tower Hamlets, and who recently struck for 10 months, were clearly much busier than me and my elderly colleague. So besieged were they with the pressing 'welfare' needs of the population of this 'deprived' London borough that they could not stand it any more and had to take industrial action to ensure both that more social workers were appointed and that they all got a lot more money. Even though, during their protracted absence, life seems to have pursued its chequered course in Tower Hamlets with massive imperviousness, the social workers seem to have achieved both objectives.

What other objectives social workers were pursuing on behalf of tax- and rate-payers before withdrawing their labour is not made clear in a report[1] commissioned by Patrick Jenkin, Secretary of State for the Social Services. The Social Work Service of the Department of Health and Social Security, an organisation which succeeded but did not replace the various existing inspectorates when Social Services Departments were invented in 1970 to subsume the existing Children's, Welfare, and Health Departments of Local Authorities, was asked to 'investigate the effect on clients of industrial action by social workers in the London Borough of Tower Hamlets' (p. 1).

No clients interviewed

The first and most fundamental criticism that can be levelled is that, although it purports to examine the effect on *clients* of the strike, the report gives no indication that any client was asked for a view. It may be that there were insuperable difficulties in obtaining the views of clients (there have usually been such difficulties in social work-controlled investigation into social work functioning). If this was indeed so, a mention of the difficulties might have prevented the inexperienced reader from speculating whether it was thought more politic to gain reports of clients' reactions from other professionals, preferably other social workers. After a long and disillusioning exposure to social work research conducted by social

---

[1] *The Effect on Clients of Industrial Action by Social Workers in the London Borough of Tower Hamlets: An Investigation*, a report by the London Region Social Work Service of the Department of Health and Social Security, 1980.

workers, I attribute no such Machiavellian sophistication. Recalling how Professors Phyllida Parsloe and Olive Stevenson spent £128,000[2] of public money on a project whose original terms of reference were 'The Task of the Fieldworker in Local Authority Social Services Departments, some implications for training', transmuted their findings without a blush into *Social Service Teams. The Practitioners' View*,[3] and wrote a lengthy book about how social workers' jobs suited or did not suit them, and how they got on with each other in 'the area team world', showing no concern about whether or how these narcissistic activities benefited the clients; recalling that no-one found this strange, and indeed that the report was greeted with enthusiasm; recalling these and other kindred matters, I attribute the Social Work Service report's failure to see any clients as yet another demonstration of the fact that many of those influential in the profession see social work with a bizarre kind of innocence as an activity existing for the benefit of the practitioners. Arrogant? I do not think so; stupid and pathetic rather, because so doomed in the long (or, perhaps, post-Thatcher, shorter) term.

Haphazard interviewing

I do not know whether Anne Howard and John Briers, a facsimile of whose signatures (the *personal* touch) appears at the end of the report, are trained social workers, but from the general disorder of their thinking and organisation, and give-away remarks like

> 'the boy *presented well* and magistrates were very concerned about the possibility that he was seriously disturbed and that his offences (theft and attempted breaking and entering) represented *a cry for help*.' (My italics.) (p. 21),

I guess they are. If so, they may well interpret their function as members of the Social Work Service as 'enabling' rather than, or as well as, an inspectorial one. Given the importance in what passes for theory in social work training of 'non-judgemental' attitudes, they would probably consider the collection of routine information which would help them to assess 'outcomes' rather than participate in 'process' unnecessary and possibly destructive to the formation of

---

[2] An under-estimate, since costs of printing and publishing are deemed to be covered by sales, and it is highly unlikely that anyone not publicly funded will actually buy the work.

[3] Department of Health and Social Security, 1978.

'relationships'. They erect no such tiresome barriers to relationship in the report. Like Parsloe and Stevenson they wander about seeing people apparently as the mood takes them, a doctor here, a headmistress of an infants' school there, a senior nursing officer here, the Assistant Regional Controller of the DHSS there, and so on. All we are told about the way informants were selected is:

> 'It was clear that . . . we needed to make contact with as many as possible of those individuals and agencies who would have relevant information about the effects of the industrial action by social workers. A range of such informants was quickly identified and others were added as the investigation proceeded.' (p. 1)

In one area the coverage was indeed exhaustive: the staff of the social services department.

> 'We consulted the Director, the Assistant Director (Operations), the four Group Controllers with responsibility for social work services, residential management, domiciliary and management services and Principal Officers and Advisers.
>
> We interviewed representatives of each of the specialist sections within the Department which deal with social work in hospitals, court work, intermediate treatment, foster care and adoptions, children in long-term care, and homeless families.
>
> All seven area officers were visited by appointment and all the Principal Area Officers in post were seen as were the majority of senior social workers and social workers. Most social workers were seen in one or more groups in each area office but some accepted the offer to be interviewed individually. We saw the Group Principal Hospital Social Worker, Principal Social Workers and representatives of the hospital social workers and support staff.' (p. 2)

Residential and other workers were also consulted, but with irritating imprecision the authors fail to tell us whether the 'heads of five residential homes for the elderly, heads of day care establishments including day nurseries, day centres for the elderly and handicapped and adult training centres and home help organisers' comprised a selection or the totality of the institutions and employees. 'A wide range' of both statutory and voluntary bodies was seen, 'representatives' of the Juvenile Bureau of the Metropolitan Police, 'two chairmen of the bench and a magistrate of the Juvenile Court', and many others chosen without apparent purpose were interviewed, but of course no clients. Another example of what I earlier termed bizarre innocence occurs on page 22 where the authors make the astonishing claim:

> 'The fact that *some* consultants and *some* senior nurses were so ready to meet us and offer their views suggested that they valued the contribution of social workers.'

They add:

> 'We heard of others who hold the view that social workers are superfluous, *but they did not seek to see us*.' (My italics.)

Surely if a complete picture were being sought the investigators should have attempted to see *them*. I hope I am wrong in seeing in this the blinkered complacency that so discredits social work and social workers, and casts profound doubt on the capacity of members of the Social Work Service to undertake what they describe as 'normal monitoring work'. It is displayed again when they say:

> 'Statutory and voluntary agencies providing a variety of professional services were approached, and *almost without exception* they gave us full co-operation.' (p.1) (My italics.)

Which services, what do you mean by 'professional', and who were the exceptions? one longs to ask. Naïvely they tell us:

> 'Many took the initiative in offering their experiences and evidence of the effects of the strike on them and their clients.' (p. 1)

Perhaps, since there was no pattern in the interviewing, it was proper to accept volunteer informants, and, presumably, to give their contribution equal weight with others, but the investigators appear totally unaware that those who volunteer information may be motivated by quite other objectives than the search for truth, as indeed may those who *are* specially selected.

'Aimless and formless' collection of 'information'

Given that the selection of respondents appears to have been haphazard, to say the least, how was information obtained? 'We interviewed the great majority of informants *individually or collectively* using a *standard procedure*' (my italics). No clues are given as to what this 'standard procedure' was, nor how it was possible or why it was necessary to apply it to some 'individually' and others 'collectively'. Parsloe and Stevenson in *Social Service Teams* tell us that the questions they asked of their respondents altered as they went along and 'learned by experience', but they too fail to reveal the questions either in original or 'improved' form. A scientist who was kind enough to read both reports, incredulous that what I reported was true, remarked: 'One hopes the evidence presented is answers to questions. If neither the questions nor the answers (the raw data) are revealed, the work of the researchers is wholly useless, since their findings cannot be checked'. While it would be ingenuous to imply that the methods appropriate to numerate sciences can be transferred *tout court* to the so-called 'social sciences', it is distressing that, to put it at its kindest, these apologies for an inspectorate do not even seem to be aware of the existence of 'standard procedures' and guilelessly describe their aimless and formless meanderings. Remembering Sir Karl Popper's view that much of the trouble in the world is caused not by man's evil cleverness but by his stupid goodness, I do not believe questions were omitted because they were loaded and the researchers were ashamed of them. I think it far more likely such questions as were originally posed were abandoned as the 'research' proceeded in favour of cosy chats that did not get in the way of 'relationships', and the researchers hardly realised it.

'Comparative figures'?

Early in the report the authors tell us they were seeking 'factual information, including comparative figures', and that 'statutory and voluntary organisations . . . produced facts and figures wherever possible, at some considerable time and cost to themselves' (p. 3). The only hard information justifying the description 'comparative figures' in the whole report was obtained from records kept by the 'senior management' who manned an emergency desk during the strike, and from existing DHSS statistics. It is that, during the strike, 182 children were admitted to residential care compared with 307 in the comparable period in 1977-78, and 194 as opposed to 298 elderly people were admitted to homes. It did not need a full-scale inquiry to reveal this information. Nor is it certain that it reveals a deterioration in service. Another 'statistic' tells us that of an establishment of 245 field social workers there were 25 vacancies when the strike began, 93 a year later.

Given the tale of ineptitude that the report unwittingly unfolds, I for one do not find this a cause for mourning. Supposing, as one must, that the Social Work Service comprises people eminent in their profession, specially gifted and experienced, and if this report is the best they can do, it would be no loss if the whole shebang were gently obliterated. If social workers are even less competent than these preposterous investigators, Tower Hamlets and the rest of the country would be better off without any of them, a suspicion I have long harboured.

Report fundamentally flawed by unscientific method

It would be dishonest to close having left the impression that, flawed as the methods used by the Social Work Service were, I or any of my colleagues could have done much better. Other papers have made clear the limitations of research as a guide to policy-making, and have talked of the nebulous but emotionally appealing concept of 'need' for services to which welfare professionals so constantly appeal. As my scientific colleague dismissively remarked on hearing that a course entitled 'Need, normative and felt', was to be included in a new degree package designed to attract students to the social sciences: 'When social scientists say "need" they mean what they think I ought to want, so that they can tax me to pay them to give it to me'. This inquiry could hardly be dignified by the name 'research', since it was a retrospective attempt to assess a service whose objectives were unclear and the 'need' for which was unestablished. However sophisticated the methods employed to tackle it, it would be naïve to expect an answer to an unformulated question. It would be equally naïve to expect effective inspection of a service staffed by those who, as Dr Anderson so succinctly put it, confuse their intentions with their abilities, especially when the inspectorate comprises people whose training may have clothed the same defect in bogus theoretical trappings.

III. MORE MEANS WORSE?

However unable social workers may be to tell us what they are doing and why it is useful, they can be relied on to react to any situation that catches public attention with a plea for more of it. I can do no better than close with a report from the social workers' house journal, *Social Work Today*. It concerns a subject on which none of us can be indifferent, child abuse, in this case leading to the death of a child. But let us not equate scepticism about the ability of social workers (or possibly anyone else) to prevent this evil with callousness about infant suffering. If I thought it even probable that the appointments outlined below would lessen the risk of injury to a single helpless innocent I would be happy to fund them from my own pocket, as I am sure the rest of us would be. My considered judgement is that to augment an already top-heavy service in the way suggested will serve only to inflate the pretentions of those who have yet to demonstrate competence. Social workers do not agree, as this extract indicates:

> '*Child death case may lead to more staff*[8]
>
> Three additional deputy area directors may be appointed and some 74 social work posts regraded, in the wake of the Carly Taylor case.
>
> Terry Smith, deputy director of Leicestershire social services department, said:
>
> "Recommendations for deputy directors in three areas of the city and the regrading of all social workers to enable them to be paid more money for working in an area with particular social problems, have passed the social services sub-committee stage. As the proposals obviously involve more cash, the policy and resources committee must give their final approval."
>
> Other recommendations include the appointment of a principal assistant who would be responsible for reviewing the work of area and hospital teams. The proposals follow a series of recommendations made by Dorothy Edwards, director of Leicestershire social services, following the death of Carly Taylor, the 13-month-old baby who died in 1978 three months after being returned to her parents by social workers.
>
> The report pointed to the shortage of manpower and resources leading to pressures on social work staff and recommended the appointment of eight additional social workers to ensure the proper supervision of non-accidental injury cases.
>
> Dorothy Edwards also called for six more deputy area directors and two staff at principal officer level. The social services sub-committee have recommended that these additional proposals should be kept under active review although lack of resources means it is doubtful that all of them will be carried out in the near future.'

Or, one might add, if the rest of us get better at spotting those who confuse their intentions with their abilities, ever . . .

---

[8] *Social Work Today*, Vol. II, No. 33, 29 April 1980.

5. Three Fallacies: Ideological Error in Local Government Thinking

DAVID MARSLAND

Brunel University

The Author

DAVID MARSLAND: Senior Lecturer in Sociology, Brunel University, and Director of the South East Regional Unit for the Youth Service. Executive Committee member of the British Sociological Association. Author of *Sociological explorations in the service of youth* (1978), and, with Michael Day, *Black kids, white kids: what hope?* (1979).

I. INTRODUCTION

My task is to develop *three* of the themes indicated by Dr Digby Anderson. I have construed these themes as three fallacies. What I attempt here is to analyse three fundamental misconceptions prevalent in the discourse and practice of local-state welfare functionaries. I do not mean to argue that these fallacies only find expression in local government. They are sadly evident and influential in the central state apparatus, in the media, throughout the educational system, and indeed wherever thought runs shallow, honesty is rare, and fear is paramount. For good reason, however, the influence of these fallacious beliefs is particularly strong in the *local* welfare state.[1]

II. A TRINITY OF ERRORS

First I shall name these fallacies and briefly define them. On this basis I will go on to explore their effects through concrete illustrations.

(i) *Mass-production fallacy*

The first is what I shall call the *mass-production fallacy*. By this I mean the belief that local provision of services for people:

(a) is properly designed for an undifferentiated market—the mass, the people;

(b) is to be maximised rather than minimised—mass production indeed;

(c) is to be monopolised by those who know best, i.e. local state officials;

[1] Here and in the body of my paper, it may appear that I am being less than just to many hard-working, intelligent, and honest local government officers and council members. To counter this impression, I wish to state explicitly that (a) in terms of persons and personalities I am talking about a small minority, and (b) ideological error can persist regardless of individual persons and personalities because of structural causes.

(d) can operate efficiently with a manifestly naïve and over-simplified model of people.

This fallacy defines people as ciphers and pawns. It is profoundly anti-democratic, not least when it is brandished by those who most stridently claim to be acting 'democratically'. It is the crop reaped in local government from the seeds sown by Fabianism. Even conservatives seem incapable today of recognising that the municipal Goliath, like the biblical Goliath, is both a deformity and less than invincible.[2]

(ii) *Immoralist fallacy*

The second fallacy I shall call *the immoralist fallacy.* This is the belief that either:

(a) decisions involving a choice between basic values can be avoided, with moral decisions replaced by scientific, technical, or operational analyses;

or alternatively—and the worst cases use both alternatives to suit their convenience—

(b) moral decisions are self-evident and unarguable.

In both, real choice is denied to ordinary people altogether. Shallow self-righteousness and imperialistic expertise parade through local government as arrogantly and as unchallenged as the Kray brothers through Bethnal Green.[3]

(iii) *Marxist fallacy*

The third fallacy already has a name. It is *the marxist fallacy.* Its contemporary power is, however, so great that non-marxists and even self-professed anti-marxists in both the major parties are unwittingly its everyday exponents.

The marxist fallacy is the belief that:

(a) We are struggling towards liberation and progress out of benighted enslavement and poverty.

(b) The necessary mechanism for social development is increased

[2] Even those who would not accept the 'minimalist' conception of local government which I argue for here ought at least to accept that no service should be undertaken unless it has been shown: (a) that its objectives are fully clarified; (b) that it is necessary; (c) that it is operating efficiently and effectively.

[3] Never before, it might be said, have so many known what was good for so many others.

political control—since private citizens and free associations are not to be trusted with the delicate susceptibilities of the people. Only the state, locally and especially centrally, can deliver the goods of democratic progress.

This third fallacy is peculiarly difficult to rebut at a time when the only legitimate McCarthyites are fellow-travellers, and when marxist and quasi-marxist interpretations of our history and our social policy have been raised to an almost official status.[4]

III. THE FALLACIES AT WORK IN ONE SMALL SERVICE

I shall now examine some concrete illustrations of these three destructive fallacies at work, starting with an example of each in the local government Youth Service. I choose the Youth Service first, not because it is peculiarly prone to these errors—I believe it is in practice a rather sensible, modest, and clear-headed organisation compared certainly with education, social services, or most others—but simply because it is a field I have done detailed work in, know well, and respect.[5]

One of the main advantages of the Youth Service over many others is that it is highly professional in the best sense of the word, and yet the vast majority of its personnel are part-time and volunteers. Moreover it is constituted as a partnership between the statutory and voluntary wings. Historically and ideologically, voluntarism in both these senses is of the essence of the Youth Service. More recently, and increasingly, however, the first fallacy in one of its forms is disrupting this sensible pattern. The voluntary agencies are increasingly squeezed out. Where there are cut-backs, it is the part-time volunteers who are being let go. Furthermore, this newly dominant approach is in danger of being imposed on users and

[4] The rhetoric of marxist historians is increasingly triumphant. It cannot be dismissed as mere academic 'stuff'. Every policy decision implies and rests on an understanding of history. Non-marxist policies cannot arise out of marxist history.

[5] D. Marsland, *Sociological explorations in the service of youth,* National Youth Bureau, 1978, and 'Novelty, ideology, and reorganisation', Chapter 2 in D.C. Anderson (ed.), *The Ignorance of Intervention*, Croom Helm, 1980.

consumers without adequate research or proper consultation. They are construed simply as material for a local authority machine.[6]

The immoralist fallacy is most in evidence in the Youth Service in sexual and political education and in work with what used to be called delinquents. In each of these activities I see increasing signs of tendencies among youth workers, encouraged by some of the initial training agencies, to opt out of a proper professional concern with careful moral thinking on their own part or with necessary consultation with families. The 'right line' on sex, politics and responsibility for wrongdoing is taken for granted, continually reinforced in everyday conversation and interaction in networks of colleagues, and trendily 'left-liberal'.[7]

This of course leads directly to the marxist fallacy. Youth workers have traditionally been politically innocent, even to a naïve degree, compared with some sectors of the newer professions. This condition is perhaps the main reason, combined with the under-development of intellectual work as such, for the easy ride marxist ideas have been having lately. Increasingly we are encouraged to see the primary objective of youth work, not as helping individual young people in particular concrete ways, but as contributing to social change. Nor is there much ambivalence in these quarters (in the training agencies and in the influential baggage-train of journalists, consultants, and intellectual hangers-on which trails behind all the weaker professions nowadays) about what social change means. Nothing counts as social change except rapid movement towards an 'egalitarian society'.[8] Even senior youth officers, chairmen of Youth Service committees regardless of party, and inspectors seem to have let themselves get caught up in this sort of nonsense.

Antidote to the fallacies: clarity, firmness, accountability

The only antidote to it, as to the other fallacies, in the Youth Service as elsewhere, is clarity and toughness about purposes, open dis-

[6] J. Eggleston, *Adolescence and Community*, Edward Arnold, 1976; A. J. Jeffs, *The Youth Service and Young People*, Routledge & Kegan Paul, 1979.

[7] Examples can be found in the journal *Youth in Society*, published by the National Youth Bureau, *passim*. I have examined aspects of these issues in 'Sociological research on youth and adolescence', in A. Hartnett (ed.), *Educational Studies and the Social Sciences*, Heinemann, 1980.

[8] S. Butters and S. Newell, *Realities of Training*, Department of Education and Science, 1977. This phenomenon has gone much further in school teaching, social work, and probation.

cussion, and proper decision-making in accountable channels. In particular, the democratically elected and appointed committees and sub-committees of Education Departments, the management committees of Youth Service agencies, and working groups of local government officers with policy responsibilities have work to do.

Of this there is, sadly, little sign at present. The recent Commission for Racial Equality report on young blacks, for example, in its analysis of the Youth Service, works up all three fallacies almost to a pitch of hysteria.[9]

IV. THE FALLACIES AT LARGE

Systematic analysis demands a thorough researching of all the sectors of local welfare to explore the extent to which these fallacies are affecting local government as a whole. All I can manage here is a preliminary step in that direction—an indication of some of the spheres in which these mistakes are indubitably happening.

A second trinity

I have chosen three topics. But these are merely three out of many, and perhaps not the worst. First, 'Community Education', a noise which will, I suspect, ring in our ears these next five years as harshly and as hollowly as 'industrial democracy' or 'affirmative action'. Secondly, supervision, an increasingly fashionable idea designed to avoid and sabotage accountability and responsibility. Thirdly, the concept of 'area' as it is used in social services, housing, planning, transportation and community development.

(i) '*Community Education*'

This idea is scarcely less vacuous and formless as a concept than that of community itself, on which it rests. It means a variety of different things in different places, and even at different times in the same place. In some local education authorities it is used to define and to rationalise the objectives and structure of the local education system as a whole.[10] In these cases it incorporates nurseries, schools, colleges,

[9] *The Fire Next Time*, C.R.E., 1980.

[10] The Inner London Education Authority (ILEA) is perhaps the best—that is to say, worst—example of this conception of community education. (cf. *An Education Service for the Whole Community*, ILEA, 1973.)

careers, adult education, etcetera, into a vast apparatus for life-long education. In others it is more modestly restricted to scrambling youth service, adult education, and perhaps leisure into a multi-purpose omelette of about the same scope and edibility as that created in social services by Seebohm.

Whatever its particular form, community education seems to have the following characteristics:

(a) It is a desperate response to manifest inadequacies in the schools, especially as these affect what we used to call 'ROSLA' ('Raising of the School Leaving Age') children.

(b) It purports to find answers to hard, challenging questions by appealing to the magic solution of the concept of 'community' and 'community needs'.

(c) Instead of looking at the scope for practical co-operation between individuals and departments, it glibly throws away the benefits of necessary specialisms—and all this for the sake of a spurious unity in thoroughly opaque purposes.

(d) It shifts educational discussion from areas where we have at least some modest practical knowledge—about teaching, the organisation of schools, the skills of youth work, for example—into a domain of pure guesswork and hope. This is a fairy-land where disorderly concepts—equality, continuing education, disadvantage, open access, minority rights—clash in a dark night of speculative theories.

'Community Education' is already instituted in many local government areas and under consideration in even more. It is riddled with the consequences of my first two fallacies, and an easy prey for marxism.

It is a prime example of the mass production fallacy. It presupposes that a singular system can cater for a huge variety of customers and clients. In its very nature it is a large-scale monopolistic system persistently devouring and incorporating and allegedly systematising its statutory neighbours and competitors and voluntary efforts. The notion of people on which it rests is both over-simplified and idealised, and hardly tested at all through any form of consumer research. It promises a bureaucracy as gross as the National Health Service.

It may seem harsher to condemn community education for immoralism, since its spokesmen tend to be 'nice folk', anxiously concerned for the general welfare—retired Chief Education Officers, for example, professors of Adult Education, and lady Conservatives. But this is precisely the danger. It takes for granted (absurdly) the moral impropriety of privacy, of contented involvement in family and work, of quiet satisfaction with the local political structure. It demands a perpetual activist involvement in every sphere of decision-making from everyone. It encourages mistrust and dissatisfaction at every level within organisations and in every sphere—except of course in relation to the developing totalitarian apparatus of community education itself. As a result, it pre-empts moral analysis by taking it as given, and prevents, while claiming to facilitate, the political action of free citizens.[11]

(ii) *Supervision*

This idea is almost a contradiction of the commonsense meaning of the word, sometimes recognised by calling it 'non-hierarchical supervision'. It is an American importation, securely established in social work, and spreading through probation and community work to more and more spheres. It represents a reaction against the negative aspects of 'bureaucracy' and the supposed inhumanity of 'management'. It insists on the professional's need for (and right to) support from fellow professionals in one's own rank, and protection against the stringencies of responsibility and accountability for one's work.[12]

It represents an important expression of the mass-production fallacy in two senses: *first,* because the system is growing and becoming taken for granted. Training agencies are encouraging the expectation of supervision; departments are under pressure to introduce it, whether or not there is any wish for it, let alone a 'need'. *Second,* because it constitutes an irrational and unnecessary duplication. Everyone is to have not only a superior but a supervisor as well. The very least of its harmful effects is that it wastes time or money or both, and the unions defend this waste aggressively.

Its immoralism is less transparent but not less real. It presupposes

---

[11] Further comment on community education is in D. Marsland, *Sociological explorations . . ., op. cit.,* Ch. 10. On the important links between immoralism and marxism, E. Kamenka, *Marxism and Ethics,* Macmillan, 1969.

[12] P. Halmos, *The Personal and the Political: Social Work and Political Action,* Hutchinson, 1978; A. Kadushin, *Supervision in Social Work,* Columbia University Press, 1976; D. Pettes, *Staff and Student Supervision,* Allen and Unwin, 1979.

—and encourages young professionals to believe—that one's superiors in the structure of a work organisation are in principle not to be trusted with important problems, personal or vocational. It rests on an assumption that 'them' and 'us' divisions are permanent and proper. It encourages divisiveness irresponsibly. Its empty moral superiority is as devoid of real moral content as the traditional resistance in mining villages to those who look like 'getting on'.

All this nonsense might seem a silly piece of relatively harmless idealism. But increasingly the practice of supervision is a primary vehicle for radicalisation. The networks and groups it constitutes too often coincide with political cells fundamentally committed not to work on behalf of clients but to 'radical change'. Indeed, the egalitarianism which underlies supervision is logically identical with that which defines the marxist fallacy. When members of the 'caring professions' strike, we can be pretty certain that the apparatus of supervision should take at least some of the responsibility.

(iii) *'Areas'*[13]

Developments in the profession and practice of planning over the past 50 years have a lot to answer for in the current difficulties. The 'area' approach is one part of this inheritance. Today its use proliferates continually from department to department of local and national government. They all divide the bigger patch up into little bits. It seems sensible, even attractive. Unfortunately, the areas designated for different purposes rarely coincide, and in any case they are usually arbitrary. Furthermore, the application of the area approach to welfare provides a rich arena for exponents of the three fallacies.

It *seems* like an antidote to the mass-production fallacy, but in practice it is a positive expression of it. For, once the population of an arbitrary patch of ground is designated as belonging to an area—for housing action, for social services provision, for urban aid largesse, and so on—*all* of the variegated people and families included within it are arbitrarily subject to whatever privileges or pains the local state machine promises. Correspondingly, all those outside it are deprived of (or saved from) them. Money that should be spent is saved, but—to an even larger degree—money is expended unnecessarily and waste is defended in terms of mock-welfare principles.

[13] R. Lees and G. Smith (eds.), *Action Research in Community Development*. Routledge & Kegan Paul, 1975.

The moral dimension of the concept is similarly problematical. 'Is' and 'ought' become altogether interchangeable in a context where an area is displayed as special, as demonstrably a problem *in toto*. Individuals and families of quite diverse types and needs find themselves inescapably stereotyped and stigmatised.

Finally, the area concept provides a ready-made basis for ideologues and their various brands of the marxist fallacy. Even more than 'communities', areas so-called provide the home-ground for action groups working against accountable democratic controls and the best efforts of elected councillors. Backed by reference to 'my area', unscrupulous leaders have available a powerful rhetorical resource in local disputes, however fictional real support for their ideas is in practice. 'The Council' comes to be branded as the villain of the piece, and properly taken, sensible decisions are overturned.

On many grounds, we should be very suspicious of this idea and its effects on local government.

V. CONCLUSION

I should have liked to have discussed further examples, especially the ideas of 'action' and 'participation', and the still expanding empires of 'community relations' and 'community development', but space is limited.

I can only suggest that if the analysis of weaknesses in the local welfare system in terms of my three fallacies is plausible, we ought to apply it urgently across the board. These false beliefs are not the whole but they are not the least of what stands in the way of sensible and radical reforms. Unless they are dealt with, the chances of local government progressing from beggary to financial independence, from being the office boy of the central state to an autonomous rôle of accountability for its own policies to its own citizens, are small indeed.

Questions and Discussion

ARTHUR SELDON: There are a number of councillors and council officials present and I feel that at this point they ought to have their say. You have heard a couple of statements in summary which must have stimulated your thought processes.

W. U. JACKSON (*Chief Executive, Kent County Council*): I accept a lot of what has been said this morning—I found it enjoyable and stimulating. I believe there is a need for the containment of local government expenditure and of public expenditure. I like to think that my own authority, Kent County Council, is doing its best to respond to that challenge, but I want to make two assertions, stemming in part from something that Dr King said and, secondly, in response to June Lait's statement.

First, if you have got to have a degree of state bureaucracy and state intervention for the provision of public services, local government is the best vehicle for providing them. It can be improved, but it is better than provision on a central basis by a central state bureaucracy. At a local level, you can have experimentation and pioneering, you can test efficiency and make comparisons between the performance of individual authorities. But there is a need to sharpen performance, and there is a need for more comparisons to be made.

Secondly, while it was great fun listening to June Lait knocking the social services, and while there is no doubt a need for reduction, we cannot duck the fact that there is a need for social services of some kind. You cannot consult a battered baby, particularly if it is dead, about whether he has missed his social service worker. Families with mentally handicapped people do need support, and the only way in which this support can be provided at present, in the absence of adequate voluntary effort, is through local government services. Fine down your local government services by all means, but do not imagine you can do without them altogether.

GERALD FROST (*Centre for Policy Studies*): June Lait concluded that we could probably do without social workers completely, and that they have really no effect. That strikes me as being distressingly moderate. It seems to me that there is a great deal of harm done by social workers. The Conservatives amongst us know that society is held together by bonds of allegiance, affection, tradition. Much social work erodes those bonds and has nothing really adequate or satisfactory to put in their place. When social workers assume responsibility for other people's relationships, they are *taking away* responsibility. People may well question why they should look after their elderly relatives or the sick and infirm people they know in the community when there are others who will be paid for doing that duty.

I speak with some personal bias against social work because my wife was a social worker for five years. Often quite late at night, about two or three o'clock in the morning, we would receive telephone calls from people asking: 'Will you take our parents into care?' or 'Will you come and settle our domestic dispute?' Since we were sometimes engaged in these ourselves, this was extremely irksome—it would have made just as much sense to ask them to come round and solve our problem as it did for my wife to go round and solve theirs. That is not a frivolous point. It does seem to me that if we allow paid officials, even though they claim to be caring, to assume responsibilities for other people's lives, those people will stop taking responsibility, and as a result society will become more fragmented.

If I am right, the only conclusion we can reach is that social work is inherently anti-social and should be stopped forthwith in the name of a healthy society.

TYRELL BURGESS: First, I greatly sympathise with Dr Anderson about finding ways of discovering whether public policy and public institutions are doing what is required of them. I think that is an urgent job and anything he can do to help with that strikes me as being very worthwhile. I wish, however, he had actually said something about the progress that he had made in that respect, because progress is necessary.

Second, one can have a lot of justifiable fun about whether or not one would miss things if they were not there. Would it be dreadfully unfair if I suggested that it would take the country a very long time to miss the University of Swansea if by some freak of chance it should be abolished?

Should local government supply welfare? Are there some services that it is necessary for government to supply, whether roads, schools, hospitals or whatever? My view is that there are a large number of services that public bodies should supply and that, given that they are to be supplied by public bodies, it is a good thing that they should be supplied by *local* and not *national* bodies. I agree that is a pretty bland and bald statement.

If I can take the argument a bit further, I want to offer a kind of theoretical basis. There are two ways in which a pluralist society can operate. One of the mechanisms for its operation is the market, which is the way in which individual choices can be expressed and met. Another way is through democratic institutions: people can vote for accountable public bodies. My view is that each of these operates on the other, that democratic bodies have to operate with a view to the market and that it is important to check the operation of the market by the democratic expression of opinion. Sometimes the market, if it works at all, takes too long to work for it to be tolerable to many of the people affected. In the kind of society I want to live in, the market and democratic bodies interact.

I offer that as the beginning of a basis for a serious discussion about the extent to which democratic local authorities should supply public services.

SELDON: Before I ask for your questions I shall let you hear my view. I think we should repudiate uncompromisingly the notion, still taught by many who have taught it for decades, that institutions based on the ballot-box or on five-year elections are accountable to the people. The experience of the post-war years is that it is precisely because they are *not* accountable that there is a reversal in intellectual thought, opinion and research which is now rejecting the notion on which we have based our social policy for the past century: that political institutions, as Mr Wedgwood Benn is now unceasingly repeating with no evidence at all, are or can be accountable. Our political institutions have had a century of increasing trial and have been found to be *un*accountable. It is impossible for ratepayers or national voters to enforce their preferences. There is no way by which you can hold your local councillor accountable for his expenditure of your money.

That is my say. But is there an economist here who would respond to Tyrell Burgess's subdivision of the methods of running society and the supply of public goods by the market or by so-called 'accountable bodies' based on systems of voting?

GORDON RICHARDS (*Hammersmith & West London College, former Labour Councillor*): I am no longer a member of the Labour Party nor of any local authority. The notion that local authorities are accountable to their electors is really preposterous to anybody who has had experience in local government—I spent 13 years in it myself as an elected councillor.

One must first examine the method by which local councillors are selected and elected and the way they behave subsequently. The fundamental answer to Tyrell Burgess about accountability of local authorities is that most are elected on an entirely irrelevant criterion, namely, the performance of the central government. How on earth can you expect a locally-elected authority to be accountable to its electors when most of those electors voted on the basis of their assessment of the performance of the national government? You can have an entirely or a largely satisfactory Labour council of the old-fashioned sort which believes in value for money and not spending wildly in all directions. You still have some of these Labour councils in operation with the old-fashioned, hard-headed trade unionist or Labour councillor who believes that he is not there to throw public money around but to provide a responsible social service. But there is now a fundamental difference in the kind of person in the Labour Party, and in particular on local authorities, and the kind that you had a quarter of a century ago. If you are going to elect local authorities by reference to the performance of the central government, why should that local authority be accountable? You can have a very good Labour council which will be defeated merely because there is a Labour government in office which is unpopular at that time; that is what determines the election or rejection of a council, although not in safe Labour or Conservative areas.

One can forecast with almost complete precision what the election results are going to be in particular local authorities by reference to the standing of the Party that is in national government at the time. This is of fundamental and crucial importance to the argument about public accountability.

I protest against the use of the adjective 'social' as applied to so many of the activities of social services departments? They investigatc, as Gerald Frost has reminded us, things like divorce, alcoholism, delinquency and such matters. These are not *social* problems in that they originate in society: they are *personal* problems within the scope of the individual concerned.

I want to suggest, in answer to Mr Jackson from Kent County Council, that one method of phasing out a lot of the so-called social services of local government is to recruit the Salvation Army. I have not yet read about any social worker who has been present giving out soup to meths addicts under the arches at Charing Cross at three o'clock in the morning. But the Salvation Army do that. If you asked a social worker to do it he would want treble time!

DIGBY ANDERSON: Tyrell Burgess has totally failed to understand the first point I made. It is absolutely essential that we try to scrutinise the activities of local welfare officials, but it is not for *me* to show they do not work—it is for *them* to show they do and then for me to look at their answers. That is to say, it is not, with a public service, for a critic, either individually or as a group, to do large-scale independent research to make up for the lack of evaluation the officials fail to carry out.

MRS CECILIA GERRARD (*Surrey Education Committee; economist; ex-civil servant; housewife; county councillor*): I entirely repudiate Mr Richards's statement that local authority elections can be predicted totally by the voting at central government elections. This was proved in 1979 when the timing of the General Election coincided with the local authority elections. In my own area and in many others the voting was quite the reverse. The British public are inclined to hedge their bets: where they voted in a Conservative MP, they tended to vote the opposite way in the local elections.

DAVID KING: It seems to me that a lot of people here would like a local authority to do the following four things: (i) institute site-value rating in the hope that that would get rid of the slums; (ii) sell council houses and land; (iii) scrap social work; (iv) introduce fees for library books. No doubt there are others.

How do we get our local authorities to do these four things? There are two methods, and only two. One is to make the central government make them do it; the other is to do it directly. There are two big dangers with going to the central government. One is that you fall into the Marxist fallacy, described by David Marsland, of assuming that central government can deliver the goods. The second is that there is always the chance that the next Labour Government will reverse the decision. If you want your local authority to do those four things and to do them permanently, you have got a much better chance if you get central government off the local authority's back and make the local authority directly accountable to the ratepayers than if you try to get the central government to do it. Even if you can persuade your local authority to do them, there will always be places like Merthyr Tydfil which will do none of these things, but once other local authorities start showing that there are better ways of doing them, then Merthyr Tydfil will follow suit. So I would be happy to let Merthyr Tydfil do what *it* wants provided your local authority can do what *you* want.

It has been argued that councillors are not responsible to ratepayers. But why should they be when two-thirds of the money comes from the taxpayer? You cannot expect councillors to be responsible to the ratepayer if the ratepayer is not paying them for the goods they produce. In order to make them responsible, you must give them financial autonomy. It is not surprising that councillors are voted in on national issues. Local authorities have no power, since most of the decisions are taken by central politicians, and most of the finance is provided by central government.

So how do you expect local elections to be based on local issues? If you want your local authority to do the things I have listed you have got a much better chance if your local authority has had to make its own decisions with the electorate in mind than if it is a pawn of central government.

ALFRED SHERMAN: Tyrell Burgess seems to confuse public and private goods. Education is not a public good any more than public transport or housing, because they are consumed privately. A public good is like police and street lighting—they really provide welfare. The greatest welfare people in this country want is freedom from crime.

Should local government supply welfare? At the moment it supplies welfare because it is a national industry, mainly benefiting the people working in it. Most councillors, with a few honourable exceptions, are of low intellectual quality and are in it for what they can get. The socialists mainly want money out of it—they draw a heavy attendance allowance and expenses; the Conservatives—and this includes my own local authority, Kensington-Chelsea—are in it largely to bolster their egos and their inadequacy. I would argue that local authority has gone so far that the only cure is to abolish it.

SIR JOHN GRUGEON: What absolute rubbish! People give up time and talent and waste their own career time to serve local government, running budgets of hundreds of millions of pounds and barely getting enough money to pay for the car to get to wherever they have to go to perform their duties. I am running a budget of £500 million; anybody in industry or commerce would be getting about £70,000 a year and a share of the profits and various other fringe benefits in an era of high taxation. Furthermore, there are hundreds of people who sacrifice all their spare time, because you are never free in local government.

SELDON: A gallant statement but I imagine some of you might differ.

DAVID MARSLAND: First, I have no doubt that a lot of things presently done by the central state should be done instead by local authorities. We should pursue, despite the difficulties, whatever mechanisms will make local government responsible and accountable for providing such services as may be necessary. Given that, however, we must begin to think in the open about what does need to be provided by government and what is unnecessary, wasteful or harmful. Many of our problems may have been caused by errors in the way these things have been structured. How we resolve them has got to be thought through and a better balance struck between provision by state officials of whatever sort, by voluntary workers of various sorts, and by the opening out of some sectors to the market.

Tyrell Burgess's intervention exemplified a most dangerous confusion —that the political mechanism should be seen only as an antidote to the market. In many ways both mechanisms should and could be working together. If we think of the state only as a way of correcting 'bad markets', we shall continue to make the same mistakes that we have made since the beginning of this century.

JUNE LAIT: In answer to Mr Jackson, 'battered babies' is the standard defence of social work. But what are we going to *do* about them? Let me quote you the case of a little girl called Carlie Taylor, who was murdered by her parents after being returned to their care by social workers. This is what the report quoted in the social workers' journal, *Social Work Today*, proposed doing about this little girl:

> 'Three additional deputy area directors may be appointed and some 74 social work posts regraded'—that means more pay—'in the wake of the Carlie Taylor case. Tim Smith, Deputy Director of Leicestershire Social Services Department said: "Recommendations for deputy

> directors in three areas of the city and the regrading of all social workers to enable them to be paid more money for working in an area with particular social problems have passed the Social Services Sub-Committee stage." As the proposals obviously involve more cash, the Policy and Resources Committee must give their final approval. Other recommendations include the appointment of a Principal Assistant who would be responsible for reviewing the work of area and hospital teams.'

And it goes on to advocate massive further expenditure.

I happen to be both a mother and recently a grandmother, and I care terribly about battered babies, as I am sure we all do. If I thought that a structure like this, however costly to that authority, would save one single innocent from being battered, I would dip into my own pocket. You could rattle a tin and I would put in as much money as I could spare for such a cause. But I do not believe we know how to cure this problem, and I do not believe the appointment of hundreds more social workers does anything to affect it whatsoever. I do not know what would; indeed, none of us does if we are honest, except individual, intimate concern with one's friends, one's neighbours, one's children.

I would also like to say to Tyrell Burgess, that I am quite used to being attacked personally in the way that he was beginning to do in his remarks about the abolition of Swansea University. If you have read much that I have written, you will know that I believe that most universities expanded in a highly irrational fashion in the 'sixties and that most of them need cutting down to size and quite a few abolishing. I hope that if Swansea came under this axe, and it might well do so, I should not defend it on the grounds that it paid my salary. I think most people do tend to think this if they belong to an institution. I wonder whether Mr Burgess would go along with the abolition of his present employment if it meant the loss of his income. I think I would try to do so, and find a way of making a living uncorrupted by state subsidy.

PART IV
WHITEHALL LEVIATHAN: WILL LOCAL GOVERNMENT SURVIVE?

6. How Far Can Local Government Act With Real Autonomy if Financed Largely by Central Government?

Address by

SIR JOHN GRUGEON

Leader, Kent County Council

Chairman: ***Lord Harris of High Cross***

The Author

SIR JOHN GRUGEON: Leader, Majority Party, Kent County Council, since 1974. Chairman, Policy Committee, Association of County Councils, since 1976. Educated at Epsom Grammar School and Sandhurst. After a career in the Army, he joined the Save and Prosper Group in 1960. Member, Medway Ports Authority, since 1977. Chairman, Finance Committee, Association of County Councils, 1976-79.

Session Chairman

LORD HARRIS OF HIGH CROSS is General Director of the Institute of Economic Affairs.

Introducing Sir John Grugeon, **Lord Harris of High Cross,** *the Session Chairman, said:*

You have already had a whiff of controversey and should now stand by for more. The Seminar was planned by Arthur Seldon as a kind of 'academic sandwich', with thick slabs of economists and sociologists in both morning and afternoon, and in between this slice of red meat in the form of a real live practitioner of local government. Sir John Grugeon is no stranger to the writings of the IEA, and as leader of Kent County Council has tried to inject more awareness than most of market realities into the processes of local government. I now invite him to indicate directions in which the Town Hall can better serve those it is supposed to represent.

SIR JOHN GRUGEON:

My address today concerns local autonomy. In particular I will concern myself with two questions: Why local autonomy?—and local autonomy to do what?

'Local government' is an often-used term, but I believe it is seldom fully understood. We in Britain tend to see it in terms of democratically elected local authorities, with their own local source of finance—the rates. It is these two factors which give local authorities such as my own County their autonomy, freedom and accountability.

There are, however, many other units of government providing services to the public at local level like the local Employment and Social Security Offices; the Regional, Area and District Health Authorities; and the currently bruised and battered Water Authorities.

These provide local services in a fundamentally different way from local authorities in that they are not elected by the public they serve. Therefore they cannot be held accountable by the local electorate. This state of affairs exists even though the organisations concerned might have a considerable range of discretionary and executive powers and, in the case of Water Authorities, the ability to levy a rate.

Public services 'essentially undemocratic'

Few of us here today would regard this situation as desirable. It is essentially undemocratic that major services such as health and water should be provided by bodies which are not directly accountable to, and influenced by, the public they are intended to serve.

The arguments for public services being provided by organisations directly accountable to the electorate are obvious. (It does not need to be spelt out in detail to an audience as illustrious as this.) We have only to look at the tiers of administration in the health service and to reflect on the consequences: unproductive work generated by the simple need to communicate and agree; a high cost, not only in terms of money but also in time and morale.

Local accountability also means sensitivity to local issues and local needs. It means that services are provided at a level which is appropriate to local needs. It means local diversity; it means local choice. But perhaps most important, it means the best chance we have of getting public services on an efficient and economic basis: first, because they can be tailored to local needs; and, secondly, because the electorate can exercise an influence through the ballot box if they are not.

In services such as education, local needs and requirements will differ widely between areas. It is essential that policy decisions are taken locally, in order that this diversity of need can be reflected. The implementation of a uniform national education policy may not be in the best interests of our children or of the community, and can even lead to the destruction of education by degrees.

At the same time, however, we have to ensure minimum standards are achieved; whilst pressure also exists to ensure the best use of scarce resources.

Central government, of course, has responsibility for the national economy. In this respect it must be concerned about the size of the local government budget. This is not only because local government expenditure this year will account for about one-ninth of the GDP; it is also because, through grants, central government will meet 61 per cent of all local government expenditure.

Local government, of course, cannot do it all. Central government has overall responsibility for how services are developed. It is only right that government should decide, in general terms, how the education service or roads programme should be developing. People see these matters as national issues, and vote accordingly in general elections.

But central government does far more than establish general policy and economic guidelines for individual local authorities to follow. It has a wide range of weapons in its armoury to exert influence on local authorities. We all know of the dangers of rule through Statutory Instrument and Circular.

The relationship between central and local government is not one way. Both sides have an input; both sides listen to what the other has to say. But the balance has steadily shifted towards the centre in recent years, and it is still moving that way.

There is an ever-increasing danger that local authorities will, over time, become no more than local agents for central government.

Local Government Planning and Land (No. 2) Bill: 'steady march towards centralisation'

It is in the light of this steady march towards centralisation that all three local authority associations in England have come together in recent months to oppose certain clauses in the Government's Local Planning and Land (No. 2) Bill.

The block grant and capital expenditure proposals contained in that Bill will give to the Secretary of State a discretionary power to penalise individual local authorities which he sees as over-spending. The Bill will give to the Secretary of State the power to concern himself with the budget of individual authorities, as against the totality of local government expenditure.

This will be a fundamental shift in principle. It could in effect make the Secretary of State accountable for the size and content of each authority's budget to Parliament.

We can all imagine the effect this shift in emphasis will have in Whitehall. It may very well result in rule by civil servants. And of course more civil servants will have to be recruited to ensure that local authorities are monitored effectively, and more local government officers to respond to the resultant demands for more information.

The grave danger is that of conformity. National policies will be imposed on local authorities with little regard to the local situation.

It is no good saying that is not what is intended: I do not doubt the Secretary of State when he says less government, not more; I do *not* question his belief in freedom and accountability. But I *do* challenge his view that his formula for block grant and capital controls will have the effect of enhancing choice and freedom.

It is not by their words we should judge them but by their decisions, and the powers in the Local Government Bill could—I am not saying will, but *could*—lead to a situation where a cautious authority was under pressure to increase its level of expenditure to conform to a standard that is related to a Marsham Street view of how much should be spent.

The argument for local accountability and autonomy, therefore, is the argument against centralisation. But it is not enough to say centralisation is bad, therefore democratic local government must be good. Local authorities can get it just as wrong as central government.

If Kent gets it wrong, however, the community in Kent suffers; when central government gets it wrong the nation suffers. I hope this audience will agree that one of our biggest problems has been too many governments getting too much wrong over too long a time.

Less government, more choice

Local government is about difference, about choice. We are at a critical stage in the development of the art of government in this country and particularly on an approach to managing the economy. Too much government or too little; a monetarist approach, a Keynesian line; the welfare state, the freedom state. I suspect I know where most of this audience stands.

My personal belief is that we should play hard for less government. Bureaucracy is an expensive way of providing services, particularly when there are two or three or four tiers of bureaucracy involved. I am thinking in particular of the Regional, Area and District Health Authorities, with Community Health Councils thrown in for good measure.

I am confident, for reasons I will give later, that I am right. I believe that local government has a significant contribution to make to the reduction in the level of government expenditure. Confident as I am, believing as I do, I cannot be absolutely cast-iron certain. This is, for me, one of the most important arguments for local autonomy. It is the best means available to us as a community of testing new approaches to the provision of services.

Local government has been very effective in keeping within its cash limits since they were established in 1975. However, even with cash limits, the total amount spent by public authorities is very significant.

But cash limits alone will not make local authorities sensitive to local needs. Neither will they ensure economy and good value in the provision of services. New techniques need to be developed, existing ones expanded. We should not be afraid to innovate to achieve value for money, or to increase the voice of the local people in our affairs.

In my view, adjustments in public expenditure at local level should not be left to the Secretary of State, or his civil servants in Marsham Street. They should be determined by reference to local circumstances, and the needs and wishes of the public. Of course we must abide by cash limits. But the best people to take the necessary decisions are the accountable *local* members.

There is no complete answer to the problems we face, and we are not helped in this difficult period by the absence of a sensible government line on the possibility of a local income tax. However, there are various avenues that could be explored—indeed, in the present financial situation it is necessary that we do explore them.

The proposals I shall outline here today are different ways in which local authorities could become both more sensitive to local opinion, and also more economical in their service provision. To many members of the audience, they may appear radical. To some, they may appear impractical. I am not saying that all these ways are right, or best. But there may be areas where local authorities can push out the frontiers of government. They have done so in the past. With vision—and courage—they should be able to do so again.

There are three areas I wish to explore: charging for services; contracting services out to outside agencies; and indeed the possibility of authorities actually withdrawing altogether from some of the more irrelevant services which are provided at present. These I shall deal with in reverse order.

Before I do, however, and, more importantly, before local authorities make any decisions on these matters, it will be necessary to have a long hard think about the problems involved. I do not doubt that there will be many.

It would be wrong for local authorities to approach these matters in an incremental way. A thoroughgoing review of policy is required, not piecemeal changes.

At some stage in the resource allocation process, members should be asking themselves these questions:

1. Should we be providing this service at all?

2. Are we providing this service in the best possible way?

In answering the first question members will need not only to examine thoroughly the base of their budget but also to utilise fully their corporate management machinery to review all aspects of policy and practice. Historic reasons for undertaking certain activities should be tested.

Ideally, each area of budget expenditure should be examined each year, and justified afresh, as if new. The professional term for this approach is, I believe, 'zero-based budgeting'. I appreciate that there are technical reasons which prevent the complete adoption of this approach—but the principle is a starting point.

Opportunity costs of local services

Members should also make themselves aware of the opportunity costs of different policies. They should always be asking themselves whether there would be a better return for the public if the monies being spent on each particular service were used to provide that service in a different way or, indeed, if they were spent elsewhere.

We as Members should never be afraid to face up to the possibility that the public could be better served if particular functions were to be undertaken in the private sector. We must examine whether there are any areas of service provision from which the local authority could withdraw, leaving market forces to fill the void, if any.

Consumer Advice Centres spring to mind. So do car maintenance classes. Are the public better served by having these facilities provided by their local authority? Indeed, is there any demand for these services at all?

For those services which the local authority must continue to provide, there is no basic reason why they should always be provided directly by the authority itself. The possibility should be explored of contracting work out to the private sector, where it may be possible for it to be provided on a more cost-effective basis. *We must get out of the habit of thinking that council workmen are the only people capable of filling a hole in the road.*

This contracting-out of functions would not be a new development. There can be few, if any, local authorities which use their own direct labour forces to do all their own building work.

The point can be pursued further. Private companies, for example, might be able to collect and/or dispose of our refuse; design buildings and roads; programme computers; and provide a home-help service more efficiently than a local authority itself, and at a lower cost.

I am not saying that local authorities should hive off all their functions as a matter of course. What I am saying is that our minds should be open to these possibilities. If a private organisation can provide a better service to the public, at a lower cost, it is surely our duty to our ratepayers to ask them to do so.

There will, of course, be some services to be provided which cannot be contracted out. It is in relation to these that the local authority should review its charges policy. The object should be to root out hidden subsidies, and to ensure that the full cost of the service being provided is fully appreciated.

Two-pronged approach on charges

The approach on charges should be two-pronged. First, the case for providing services free of charge should be examined. And, secondly, authorities should examine the charges that exist at present to see whether they are set at a level appropriate both to the market in question, and to the level of usage.

In the library service, what is the reason for Agatha Christie on the rates? With regard to refuse collection, consider the careful ratepayer who burns his own rubbish, is ecologically conscious about the wasteful packaging he purchases, carefully puts aside his newspapers for the local scout group, and places his empty bottles in a bottle bank. Why should he pay the same 'rate' for refuse collection as his careless neighbour?

Looking around the room, I can see that some of you have been surprised—perhaps shocked—by some of the suggestions I have made so far. I am looking forward to the debate on these matters with considerable interest.

Undoubtedly there will be problems. But I believe we in local government would be wrong to close our minds to these new developments simply because they are innovatory, or appear radical. In the present financial climate local government must be looking for new —and better—ways of improving the quality and sensitivity of our services, and giving better value for money. If we cannot, our case against centralisation is weakened.

A policy for charging

For the benefit of sceptics in the audience, let us look in greater detail at just one of the suggestions I have made above: charges.

The changes I have suggested with regard to local fares, fees and charges should bring more sensitivity and value to local government by the simple expedient of giving to the consumer (the ratepayer) greater freedom of choice.

With charges imposed on a wide range of local government services, people would be able to buy only what they want, and leave services they do not want. Service that are unnecessary or not wanted would be revealed through the simple mechanism of the market—they would not be profitable.

Similarly, people would be able to choose between different forms of the same service. For example, under the Education Voucher Scheme as proposed by my County Council, Kent, people would be able to choose between alternative schools on the basis of the quality of the education provision on offer at each establishment.

An extended charging policy would in effect make the elector the customer for the services provided by an authority. It would create markets for different services and activities. The market would indicate where waste is occurring. Organisations offering poor value for money would become bankrupt.

The ratepayer all too often today is seen simply as the recipient of services. His voice is rarely heard, and when it is, it is not always heeded. With the consumer paying for his services directly, not only will better value for money be achieved but also local government will become far more sensitive to *real* local needs.

Further benefits will be gained from an expanded charges policy. First, excessive spending will be curbed. There is no more effective discipline on spending than knowing the price and having to pay it. You know how much you have to pay. You do not know how much if you pay by taxes. *The individual is much more effective in imposing his own cash limits than is the state.*

Second, paying by price requires a conscious decision to buy or not to buy. Paying by tax removes this consciousness.

Third, paying by price teaches the individual care in comparing values, forethought in using money and economy. Paying by taxes leads to waste and excess supply.

Fourth, prices enable payment for each commodity or service separately. At present, bureaucracies buy in bulk on behalf of the consumer. He should be allowed to decide how much of a service he want for himself.

Fifth, a change in charges policy will have a favourable effect on the financial position of local authorities. Revenue will be increased, therefore less of the budget will have to be met from the rates or, more significantly, government grants. A sensible charging policy is a healthy move away from financial dependence on central government.

Remove legal controls and requirements

The suggestions that I have made so far are radical. At present, it is not possible for them to be contemplated for many service areas. Local authorities are sometimes prevented by law from acting in certain ways. More usually, we are required by law to provide particular services in ways which are prescribed.

There are still too many central controls and requirements on local government. Legislative changes are needed to give local authorities real freedom to act: freedom to contract out services to the private sector if we feel the ratepayer will benefit; freedom not only to charge for particular services but also to decide on our own price level.

The present financial climate requires local authorities to make considerable reductions in public expenditure. This presents local government with a genuine challenge to review what it does and how it does it, and at what price.

The restoration of freedom of choice requires members to face up bravely to the task of fundamentally reviewing their authority's level of service provision. It also requires central government to appreciate that local authorities cannot do this unless they are given the freedom to do so. Freedom of choice at a local level requires less nannying from the centre.

There are some who—perhaps with justification in some cases—would look at local government with less than enthusiasm and might not be over-impressed with the argument that what we are after is a right to make mistakes at a local level instead of having them made for us at a national level.

I am conscious too of the fact that the pontificating local councillor can be every bit as infuriating as the rather more remote national politician. But I would suggest that one reason why local government does get a bad press, why it can generate a great degree of local heat, is because decisions can be related to local circumstances; because people do understand what is happening and, what is more, can do something about it, not only through the ballot box but also through the local pressure-group and the local press. It would not be a bad thing if the health service and the water authorities had to experience this kind of exposure and direct accountability.

With the necessary freedom to act, we in local government can then investigate new ways of solving old problems at lower cost. Think not what local government can do for you, but what you can do for local government!

Questions and Discussion

LORD HARRIS OF HIGH CROSS (*Chairman*): I do not think Sir John would mind my saying that his advocacy has drawn to some extent on *Charge,* the basic text by Arthur Seldon, which I strongly recommend.

ROBERT JONES (*Federation of Civil Engineering Contractors; local councillor*): First, I should like to ask Sir John about the imperfectibility of the present system, where large ratepayers such as commercial units have no say in the local authorities and their budgets. Secondly, since he put up a powerful case for charging as a more accountable procedure, what would happen in situations where the ordinary man in the street cannot shop elsewhere—for instance, with sewerage charges, planning charges, building control charges. What in these cases is the discipline that will stop local authorities from becoming inefficient?

SIR JOHN: I accept that industry pays more than 50 per cent of the rate revenue of my county, or any other precepting authority. But the CBI have the remedy in their own hands by permitting better managers—middle management—sabbaticals to look after industry's interests and provide management talent and expertise to local government. This country at the moment does not provide people of quality at local government level. Industry has got to get off its back and help us by providing council members.

LORD HARRIS: Should the business man have a double vote? If you have a business in a constituency, should you have a vote there as a businessman and as an individual?

SIR JOHN: I would not have a double vote. I would prefer an input of businessmen as council members because that is far more important.

As for discipline over local authorities, this is imposed through the ballot box. Further, I do not accept the idea that paying for planning has a deleterious effect. Planning costs a lot of money and many people put in frivolous applications. It is revenue through charging that will reduce the rate bill. Discipline must also be exercised over the councillor so that extra cash is not used for prestigious projects but to reduce the precept or the rate demand.

CECIL MARGOLIS (*North Yorkshire County Council*): Most of the things that Sir John spoke about I disagree with heartily. He seems to be under the impression that local government is under the direction of the elected members. This is not so. It is my contention that local government is under the direction of the bureaucrat.

SIR JOHN: I will say just one thing: the calibre of your councillors must be pretty poor. I do not believe that North Yorkshire is under the bureaucrats; Kent is not. I run the shop, not the officers, and once an officer crosses the line between policy decision and implementation—chop, chop!

DAVID KING: I take it from Sir John's earlier remarks that he is in favour of paying councillors. Is he in favour of them being full-time professionals or part-time professionals, or what?

SIR JOHN: I am in favour of responsibility allowances. I think it is absolutely ridiculous that the chairman of an Area Water Authority spending £126 million should enjoy a salary of £17,500 per annum for two days a week. As leader of Kent County Council, I run a £500 million a year business; I draw an attendance allowance and a motor mileage allowance that pay for barely half the costs I incur. I do not believe in 'petty cash politicians', but I consider we should not be left out of pocket. With inflation you cannot rely on the generosity either of business or, indeed, of the individual who has retired on a fixed income. Some councillors travel a round trip of 100 miles a day. In Kent the Chairman of Education is running a budget of £190 million and is also a university professor at the London School of Economics. I am out of pocket and that is my fault, but I want to encourage younger generations to follow me and I want to give them some incentive.

RICHARD RITCHIE (*The Selsdon Group*): Would Sir John not agree there are certain areas of the country where councillors are earning more as a result of their being councillors than they could probably get in employment? If your desire, with which I agree, is to re-institute charges and cut back local government, surely in practical terms you are much less likely to do that if you cultivate and encourage a form of professional councillor? People get many rewards from going onto local councils: if they are interested in politics, they might get political advancement and honours. It is not necessarily money they are seeking.

SIR JOHN: If the Local Government Act of 1972 failed, it failed for one reason only; but both political parties did not realise the magnitude of the challenge that was required. What was required was a better calibre of member. Now this nonsense about claims needs to be looked at very carefully. I have worked out that the maximum you can get, if you claim for everything in 222 working days, would probably bring you up to £3,000 per annum, assuming that you work over eight hours a day. I cannot accept that you can make a living out of it. If you do, you must be in the poverty trap.

GORDON RICHARDS (*Hammersmith and West London College*): Would Sir John like to comment on the fact that the number of meetings of the Greater London Council may be fortuitously increased from something like 4,000 to 5,000 per annum after the introduction of attendance allowances? Secondly, does he think that nationally negotiated wage and salary levels seriously militate against accountability in local government, given that some 73 per cent of all local government expenditure goes on wages and salaries?

SIR JOHN: I am in no position to criticise the number of meetings that the GLC have. That is a matter for locally elected members, and you as an elector in that area presumably should express your concern through your protest groups. You say 'nationally negotiated'. Do you want plant bargaining?

RICHARDS: Yes.

SIR JOHN: What would happen then in a small county like North Yorkshire or Durham? They would suffer through having to pay extraordinarily large wage claims because the larger authorities would be able to afford them. ILEA would corner the market, and the other authorities would have to follow ILEA's pattern. ILEA is currently spending possibly 12 per cent more *per capita* on its education than North Yorkshire, and 123 per cent more than West Sussex. I accept that there are problems in national negotiating machinery and until we re-organise LACOSAB (the Local Authority Conditions of Service Advisory Board) we will not get very far, but again it depends on the calibre and experience of the councillor to stand up and know something about wage negotiations. Over the last nine years we have forgotten how to negotiate wages and salaries because it has all been done for us by statutory control.

PART V
THE ECONOMICS OF POLITICS AND BUREAUCRACY IN LOCAL GOVERNMENT

7. The Economics of Bureaucracy and Local Government*

KEITH HARTLEY

University of York

*Helpful comments were offered by David Austen-Smith, John Hutton and Peter Watt; the usual disclaimers apply.

The Author

KEITH HARTLEY: Reader in Economics, University of York. Educated in Leeds and at the University of Hull. Visiting Associate Professor, University of Illinois, 1974; NATO Research Fellow, 1977-79. Author of *Problems of Economic Policy* (1977); with R. Cooper, *Export Performance and the Pressure of Demand* (1970); and 'Choices in Defence Expenditure' (*Journal of Economic Affairs*, October 1980). For the IEA he has written *A Market for Aircraft* (Hobart Paper 57, 1974); and 'Can Trade Unions Raise Real Wages?', in *Trade Unions: Public Goods or Public 'Bads'?* (IEA Readings 17, 1978).

Commentary Author

MERRIE CAVE: Lecturer in History, Hammersmith and West London College, since 1974. Active interest in local government: contested Islington local elections in 1974.

I. INTRODUCTION: THE POLICY PROBLEM

Current policy towards the public sector aims to cut expenditure and enforce cash limits, whilst improving efficiency and reducing waste. There is also a commitment to reduce public sector manpower, with an associated desire to change the labour mix by economising on the numbers of civil servants, reducing local government manpower and restricting the number of managers and administrators in the health service. Already, the Government is finding that it is costly to achieve compliance with its objectives.

How can a Cabinet, a Minister or a Council ensure that their wishes are actually implemented? Problems arise because bureaucrats are experts on the possibilities of varying output, as well as on the opportunities for factor substitution. Although some of these substitution possibilities can result in perverse outcomes for the Government or a Council (assuming that they have a clear idea of their policy objectives), they might be too costly for an individual Minister to monitor, police and eliminate completely.

Behaviour in NHS high-technology equipment buying

Consider an example of behaviour in the National Health Service (NHS) in the buying of new high-technology medical equipment. At the annual planning stage, doctors and specialist groups within hospitals will submit bids for new capital equipment (e.g. scanners; X-ray equipment; a new computerised cardiac arrest unit). Assume that it is Dr A's turn in the queue (a point requiring further analysis) and that he is allowed to acquire, say, new X-ray equipment on condition that he finances its running costs from his annual recurrent budget, which is subject to cash limits. Motivation and behaviour within this framework have some significant features:

First, no administrator can challenge a doctor's professional judgement on the need for the equipment, nor whether a cheaper item might be as effective. A typical story is that a doctor in, say, Burnley has a Japanese scanner with all the latest gadgets and his friend in Newcastle wants the same or even better!

Second, a doctor's employment contract provides no inducements to respond to economic incentives: he does not share in any savings. Indeed, any savings might accrue to rival departments within the hospital, or to the Regional Health Authority or even to the Treasury. Thus, the system creates incentives to *spend.*

Third, doctors can respond to cash limits by using their specialist knowledge and discretion to reduce output in *their preferred areas.* An obvious strategy is to reduce output in those areas which will increase a doctor's chances of obtaining a larger budget in the future. He might, for example, lengthen the queue for X-rays by making short-run economies in co-operating labour inputs (e.g. clerical staff).

Local government bureaucratic behaviour under economic restraint

Such examples are not unique to the NHS. This paper analyses the behaviour and rôle of bureaucracies in the context of current debates about the size of the public sector, especially at the local level. Bureaucracies can be defined as non-profit public sector organisations, embracing central and local government departments and state agencies.[1] Questions arise as to whether economic models offer any testable predictions about the behaviour of public bureaucracies. In particular, do the models embrace issues such as budgets, inefficiency, waste and manpower which dominate current policy?

II. THE BEHAVIOUR OF BUREAUCRACIES

Bureaucracies operate in a political market-place which also contains voters, politicians and interest-groups of consumers and producers. It is not without significance that government employees are also voters. Bureaus supply information, goods and services to the governing political party and ultimately to the local community. Such commodities are often supplied at zero or subsidised prices at the point of consumption, with voters paying indirectly through local rates and national taxes.[2] Moreover, bureaucracies are usually

[1] Bureaucratic organisations in private firms, especially large corporations, differ from those in the public sector in being subject to the discipline of the capital market (e.g. take-overs) and the ultimate objective of profitability.

[2] A. K. Maynard and D. N. King, *Rates or Prices?*, Hobart Paper 54, IEA, London, 1972.

monopoly suppliers, protected from possible public and private sector rivals through a local council's allocation of property rights and restrictions on entry. Each local government department tends to specialise (e.g. highways, housing, transport), there being no competition between various departments in the same or even different local authorities for the provision of specific services.[3] Nor are private firms invited to tender for some of the traditional functions of local government (e.g. refuse collection, municipal passenger transport services, careers advice). Competition from rival public and private organisations would provide local councils with alternative sources of information and comparative cost data. In other words, there are opportunities for extending competition which can result in beneficial, rather than wasteful, rivalry and duplication.

Bureaucrats' preference satisfaction predominates

In the current non-competitive framework, bureaucrats have opportunities for satisfying their own preferences; their employment contracts are often incompletely specified, so permitting discretionary behaviour. As a result, utility-maximising bureaucrats are likely to prefer larger budgets and discretionary expenditures which are reflected in organisational 'slack'.[4] This allows them to satisfy their preferences for such 'goods' as salary, promotion, expenditures on staff, new offices, leisure and even payments for consultancy reports which support the bureau. Indeed, local authorities have been criticised for over-staffing and operational inefficiency, especially in direct labour organisations; for the proliferation of departments and chief officer posts, and for the upgrading of posts and internal promotion following local government re-organisation; for unneces-

[3] At specified intervals (cf. ITV franchises), why not allow, say, the municipal passenger transport or fire departments to bid for the refuse collection service? Similarly, housing repairs and highways maintenance could bid for housing work, and *vice-versa*. And competitive bids could also be invited from other local authorities as well as from central government and the nationalised industries, e.g. Post Office supplying passenger services in rural areas as part of its mail collection. The aim would be to allow rivals to bid for a specified level of service and to offer tenders on alternative levels of service, so allowing a local authority opportunities for re-contracting.

[4] W. Orzechowski, 'Economic Models of Bureaucracy: Survey, Extensions and Evidence', in T. E. Borcherding (ed.), *Budgets and Bureaucrats: The Sources of Government Growth,* Duke University Press, N. Carolina, 1977.

sarily high standards in buildings and equipment; for defective incentive bonus schemes; and for a spate of new town halls.[5]

The outcome is that activities supplied by local and central government departments are likely to be 'too large', with output produced X-inefficiently. Hence the claim that bureaucracies result in waste and inefficiency. In pursuing a larger budget, however, they can create an *impression* of efficiency through, for example, over-estimating the demand for and under-estimating the costs of their preferred policies, as well as claiming that output is difficult to measure.[6] In some instances, such difficulties of measurement might simply reflect that there is nothing to measure.

Will cuts in a bureaucracy's budget reduce waste, raise efficiency and release manpower? A budget reduction is likely to reduce output, but officials and civil servants might respond by labour hoarding. They might try to protect jobs by shifting the bureaucracy's production function and offering only token manpower savings. After all, in the absence of alternative suppliers and cost yardsticks, bureaucrats can shift a department's production function at their discretion. Take the case of the general decline in students for teacher training. In the education faculty of one polytechnic, the average lecturer teaching hours per week dropped from 16 in 1977 to 8 in 1978 and total annual teaching hours fell from 9,000 to 4,000: staff were reduced from 50·7 to 47·1.[7] It might be argued, of course, that productivity increases in the public sector are low relative to the private sector and that this is a source of rising relative costs of government-supplied services.[8] But is this low productivity an inherent characteristic of the technology involved in state-supplied services, or does it reflect the behavioural framework within which bureaucrats operate? It would not be too difficult to test these alternative hypotheses by allowing extensive experiments in com-

---

[5] *Local Government Finance* (Layfield Report), Cmnd. 6453, HMSO, London, May 1976, p. 90.

[6] W. A. Niskanen, *Bureaucracy: Servant or Master?*, Hobart Paperback No. 5, IEA, London, 1973; also K. Hartley, *Problems of Economic Policy*, Allen and Unwin, London, 1977, Chs. 3 and 11.

[7] Local Government Audit Service, *Report of the Chief Inspector of Audit for the year ended March 1979,* Department of the Environment, 1979, p. 31.

[8] W. J. Baumol, 'Macroeconomics of Unbalanced Growth: The Anatomy of Urban Crisis', *American Economic Review,* Vol. LVII, No. 3, June 1967, pp. 415-426.

petitive bidding for some of the functions traditionally undertaken by local government.

III. EXAMPLES: COST ESCALATION AND CONTRACTS

A study of cost escalation on projects provides insights into the behaviour of budget-conscious bureaucracies with their incentives to under-estimate costs and over-estimate demand. Cost escalation is the relationship between initial cost estimates and actual expenditure (constant prices); the concept can be applied to time slippages and the quality of projects (excessive 'gold plating'). This subject also raises wider issues for local authorities, namely, project appraisal, the criterion for decision-making, procurement and contracting policy. Table I shows that cost escalation is extensive and is not confined to local government. (Appendix 1, p. 118, presents a case study.)

In Table I, escalation factors of 2·0 or more are not unknown, suggesting that these items are significant for the control of public expenditure. Escalation in the private sector raises different issues for efficiency: the costs and rewards of a decision are more fully thrust upon the selector with the risking of private funds, rival firms and consumers providing the policing mechanism for cost control

TABLE I

EXAMPLES OF COST ESCALATION

Project		*Cost escalation factor (at constant prices)*
Exeter Hospital		1·2
Harrogate Conference Centre		1·5+
UK Private Commercial Projects (e.g. electronics, chemicals)		1·05-1·5
Liverpool Teaching Hospital		1·9
NHS Computer Programme		2·0
London Court Conversions:	Court 1	2·1
	Court 2	3·01
	Court 3	3·22
Concorde		4·7
UK Guided Weapons		17·4

Sources: Reports of the Public Accounts Committee, HMSO, London.

on new commercial projects.[9] The political market-place lacks such controls. How can voters express their views on a specific issue such as a local authority conference centre or a new sports complex? How does a local councillor challenge a bureaucracy's information and advice on project selection? Moreover, bureaucrats and politicians generally operate in committees. Unlike private firms, such committees are not at risk and they are likely to choose risk-averse strategies (compromises): if they are wrong, the impersonal committee is to blame and the ratepayer finances the costs of failure.

'Too low' cost estimates lead to inefficiency

Unreliable cost estimates which are 'too low' can lead to inefficiency in resource use. Local councils are often required to select a project without being aware of the reliability of its cost estimates. As a result, under-estimation of costs might lead a local authority to buy 'too much' of a project which *appears* to be relatively cheap. Once started, projects are difficult to stop. Agents in the political market-place have an interest in continuation. Projects create interest-groups of architects, engineers, surveyors, contractors and unions, each with relative income gains from the continuation of the work. Such groups are likely to support the bureaucracy with a budgetary involvement in the project. And bureaucrats can readily show vote-conscious politicians that a project is in the local interest and will produce substantial 'social benefits' in the form of jobs, tourism, increased spending, the prestige of becoming a 'leading centre', all of which will, allegedly, 'secure' the long-term economic future of the area.

Economic models explain cost escalation in terms of urgency, modifications, unforeseen technical problems, contractor optimism and procurement policy (Appendix 2, p. 121). More resources might, for example, be required because of alterations and improvements, or in an effort to solve unforeseen problems (e.g. unexpected earth faults and water can affect the construction of foundations for buildings and bridges).

Interesting questions arise about the extent of competition for

---

[9] There is the danger of comparing real-world bureaucracies with ideal, rather than actual, markets: hence the claim that critics of bureaucracies often compare muddle with model. (cf. C. Brown and P. Jackson, *Public Sector Economics*, Martin Robertson, London, 1978, pp. 141-44.)

local government work and the type of contract which finances cost escalation. Once a contractor has been selected—and bureaucrats organise the extent and form of competition—cost-based contracts are unlikely to deter modifications or ambitious technical proposals. Optimism might also be reflected in the under-estimation of costs, with an income-maximising contractor trying to 'buy into' attractive new local authority programmes by offering optimistic cost, time and performance estimates, as well as prestige design proposals. It might be thought that such behaviour is likely only in non-competitive markets with cost-based contracts (e.g. high technology work).This is not so. Selective tendering and fixed-price contracts might not prevent cost escalation. There are examples of major construction projects, involving more than two years' work, where the successful bidder tendered a 'keenly competitive price' and was awarded a fixed-price contract. But such contracts contain provisions for price increases and any modifications required have to be negotiated separately and are not subject to competitive re-bidding: once the contract has been awarded, the firm becomes a monopoly contractor. There are examples of major modifications on building projects occurring two years *after* the award of a fixed-price contract which was originally expected to last 2½ years. In other words, cost escalation is predictable in the absence of a competitively-determined, *firm* fixed-price contract for a 'frozen' design.

Contractor optimism, bureaucrats and interest-groups

Contractor optimism and cost under-estimation could be reinforced by budget-conscious bureaucracies supported by interest-groups of architects, engineers and planners with a preference for new designs and for expanding the 'frontiers of technology'. To raise their budgets, bureaucracies have an incentive to under-estimate costs and exaggerate demands. This affects the way in which bureaucrats present information to local politicians.

Consider the case for a new local authority-financed building, such as an arts, conference, leisure or sports centre.[10] It can be argued

[10] The arguments presented in this section are a summary of points made in relation to various public sector projects. Ratepayers and taxpayers will be familiar with the language. Criticisms of local government officials do not, however, provide an argument favouring more involvement of central government bureaucrats.

that the project is 'vital' to prevent a city 'sliding down the league table' and that 'we must go ahead and subsidise the scheme because our rivals have done so'. References will be made to social benefits in the form of extra local spending from the new centre which, together with existing business, will give a 'substantial' total benefit: note the potential confusion between *marginal* and *total* benefits. On the cost side, estimates might be presented not on a consistent price basis but with, say, 1975 expenditures simply added to 1979 outlays. Thus, this method of procedure, together with the omission of interest charges and the neglect of life-cycle and external costs (e.g. noise, pollution), under-states the true opportunity cost of a project.

Further 'pressure' can be put on council decision-makers if the scheme is presented with a 'keenly competitive price' determined by selective tendering, providing a 'unique and final opportunity' to proceed—and, after all, if the project does not 'go ahead', all the previous expenditure will be 'wasted'. Indeed, completion for its own sake sometimes becomes a point of honour. It is not unknown for a council to argue that

> 'having embarked on a scheme and having for some time been fully committed to it, it is the Council's view that it must be carried through to its conclusion, notwithstanding the heavily inflated costs.'[11]

Whilst the arguments appear persuasive, they are often emotional, lacking economic analysis and empirical support. For example, use of the word 'vital' invites the question: Vital to whom?—and is it vital regardless of cost? The 'rivals are subsidising' argument is dubious since, if they wish to offer free gifts, a local community could respond by accepting them and specialising elsewhere. Indeed, in appraising any scheme, the likely costs and benefits of alternative projects have to be considered, including a 'do nothing' option (i.e. 'slide down the league table'). Nor is the 'substantial benefits' argument convincing in the absence of evidence showing that the benefits are larger than could be obtained from alternative uses of the resources.

At its most general, the benefits argument simply suggests that *any* new local project will have multiplier effects; by itself, it is not a

---

[11] Harrogate Borough Council, *Conference Complex: Public Issues Report No. 9*, July 1979; also *Harrogate Advertiser* (various issues).

convincing argument for choosing an arts centre rather than, say, new houses, schools or roads. As for costs, references to a 'keenly competitive price' are misleading for decision-making if the project design has not been 'frozen' and a firm fixed-price contract cannot be awarded. Arguments about cancellation are also confusing, since previous expenditures are 'sunk' costs where the sacrifices have already been incurred. Nor is cancellation necessarily 'wasteful': it may be cheaper than continuing with the project; and past expenditures can provide benefits in the form of valuable information and practical knowledge.

IV. POLICY SOLUTIONS

Policy-makers are fond of regulating private monopolies, but less attention has been given to the regulation of monopoly bureaus in the public sector. Some general policy guidelines for controlling bureaus can be suggested:

(1) *Question the role and extent of local government: What is the 'proper' business of local government?*

The standard economic case for council involvement would be that, if left to itself, the local economy will fail to provide services which the community might regard as socially desirable (e.g. police, civil defence, education, street lighting). Whilst this might provide a case for council *finance* of some socially desirable activities, it does not necessarily require council *provision*. Nor does the standard argument justify all the activities currently undertaken by local authorities: some functions are more likely to be explained by the economics of politics and bureaucracies (direct labour departments?).

(2) *Extend competition and private provision: Can services be supplied at lower costs?*

Massive opportunities exist for extended experiments with competition for services traditionally supplied by local bureaucracies (e.g. housing and road repairs, transport, refuse collection, design work). Rival firms would be encouraged to submit bids for a local authority-specified level of service, as well as for alternative levels of service. In this way, local authorities would obtain more information on costs where competition is feasible, with opportunities for re-contracting between different levels of service and between competing firms. US studies show that private firms are more efficient than public suppliers in refuse collection and fire services.[12] In the UK, further examples arise with local authority building projects, where design and construction are often separated, with the former frequently undertaken 'in-house' by the public buyer.

Two issues are involved, namely, the costs and benefits of separating designers and constructors, and the extent to which design work is competitive between the buyers' (public) design unit and private rivals. Separation of design and construction can lead to designs being selected on non-price criteria (with opportunities for discretionary behaviour), without consideration of the extra costs of alternative designs.[13] Moreover, where design work is a public monopoly, private construction firms will encounter barriers to entering the *design* market (i.e. offering to design *and* construct). As for efficiency in architectural work, one official report concluded that many public authorities were unable to demonstrate that their costs were lower than those produced by scale fees and that in some cases it would be more economic to resort to private practice.[14]

(3) *Introduce a competitive procurement policy: the case for open competition and firm fixed-price contracts.*

Proposals to extend private provision would involve local authorities in more extensive contracting. Current procurement policy is not conducive to efficiency, since it is based on the view that '. . . competition is a useful, but not necessarily essential, means to the end of achieving value for money'.[15] And it favours selective, rather than open, competition.[16] Critics claim that open competition raises the

---

[12] W. Orzechowski, *op. cit.*, pp. 252-53.

[13] Especially where public sector architects work to central government cost guidelines.

[14] National Board for Prices and Incomes, *Architects' Costs and Fees*, Cmnd. 3653, HMSO, London, May 1968, pp. 31-33. For public clients, 'in-house' design teams are used in 71 per cent of the number of construction projects: NEDO, *The public client and the construction industries*, HMSO, London, 1975, p. 34. Defence offers a further example, where Bristow has claimed that the civilianisation of helicopter pilot basic training would lead to savings of between 33 and 50 per cent: Defence and External Affairs Sub-Committee, *Joint Training of Servicemen*, HC 86, HMSO, London, November 1977, p. xxvii.

[15] NEDO, *The public client and the construction industries*, HMSO, London, 1975, p. 4.

[16] 'Open competition has little to offer over some form of selection prior to invitation to tender . . .' (*Ibid.*, p. 54.)

costs of abortive tendering for an industry and increases the risks of contractor bankruptcy. Selective competition is believed to avoid these 'disadvantages' by restricting tendering to a *limited* number of firms (typically 5 to 10) from an approved list and of known reliability. Usually, only a few firms on the approved list will be invited to bid for a contract, the aim being to ensure that, over the long run, all firms on the list have an opportunity to tender. As a result, selective competition resembles oligopoly with entry restrictions.

Its advocates claim that selective competition is 'the simplest way of demonstrating that regard has been paid to the public interest'.[17] But simplest to whom—the taxpayer, society or the officials acting as the Council's agents, and whose interpretation of the public interest is being used? What are the price implications and resource costs of selective, compared with open, competition? What about new entrants, X-efficiency and collusion? There are examples where select lists remain unchanged, so that there is neither entry nor exit.[18] There is the worry that the standard required for entry to an approved list will reflect the *bureaucrats'* preference for avoiding and minimising risks. Hence, councils are unlikely to be presented with information on the price implications of alternative risks associated with different contractors (including innovators). Nor can it be assumed that firms will be cost-minimisers when only a small group of 'approved units' are invited to tender—and the procurement agents determine the invitation list.

The existing contracts system is certainly neither without its problems nor costless to administer. Auditors have found, for example, that many fluctuations claims associated with inflation were incorrect, 'usually in the contractor's favour'; elsewhere, one contractor received over-payments of some £150,000 on contracts worth £800,000.[19] There is obviously scope for serious consideration of an alternative policy based on competitive principles.

---

[17] *Ibid.*, p. 45.

[18] *Ibid.*, p. 45.

[19] Local Government Audit Service, *Report of the Chief Inspector of Audit for the year ended March 1979*, Department of Environment, 1979, p. 4. An explanation and critique of selective tendering is in Keith Hartley, *Problems of Economic Policy*, Allen and Unwin, London, 1977, Ch. 11. Of course, if secrecy leads to inefficiency in bureaucracies, an additional policy solution might be a greater emphasis on 'open government and freedom of information'.

A competitive procurement policy would be characterised by open competition with large numbers of bidders and free entry, together with firm fixed-price contracts. The usual objection to competitively-determined firm fixed-price contracts is that unexpected cost increases might result in contractor bankruptcy. This is a strange objection since risk-bearing is the proper and specialist function of private enterprise and the existing contracts system might avoid bankruptcy only by reimbursing a firm's actual costs, *regardless of their level.*[20] Firm fixed-price contracts would also provide councils with more reliable price estimates for their projects. Perhaps the official opposition to a competitive solution reflects the attractions of a selective tendering policy to vote-sensitive councils and budget-conscious bureaucracies: it allows them opportunities for discretion in the allocation of contracts.

V. CONCLUSION

In an era of spending cuts, it has been suggested that the public sector is 'different'. Private firms can decline and be declared bankrupt, but it is claimed that there is no public sector equivalent of the Receiver. This is a strange view since governments and councils can reduce the total wage bill or terminate employment contracts. But the existence of such beliefs might reflect a basic proposition, namely, that bureaucrats (and people in general) show an infinite capacity for ingenuity. People can adjust and play any games. They respond to constraints (and their absence), and they will search for opportunities to exercise

[20] The basic issue concerns the most appropriate contractual arrangements to allow for uncertainty and the associated allocation of risks, and the costs of risk-taking between contractor and buyer. The choice is between open competition with lower prices and a belief in a higher risk of bankruptcy (what is the evidence?); whereas selective tendering is believed to reduce the risks of default, but at a higher price. Competition would also solve the problem of 'abortive tendering costs': tendering would be undertaken so long as it was expected to be socially worthwhile. Traditionally, firm fixed-price contracts have been used only for work of up to two years' duration. If this were a non-negotiable constraint (why?), large-scale projects could be 'broken-up' into smaller blocks with each stage allocated on a competitive, firm fixed-price basis.

discretion and pursue self-interest.[21] The problem is to confront bureaucracies with fixed budgets, output constraints and an employment contract with fewer opportunities for discretionary behaviour, and to provide politicians and society with information on efficient trade-offs: such are the economics of nirvana.

[21] Actions are not restricted by national boundaries. International collaboration on aerospace projects is an example. Following the cancellation of UK aircraft (e.g. TSR-2), both the RAF and the aerospace industry believe that one of the 'benefits' of joint programmes is that they are much more difficult to cancel!

APPENDIX 1

A Case Study in Local Government Spending: The Harrogate Conference Complex

The Harrogate Conference Complex provides data in an activity where there are few detailed project histories. Originally, there were two elements in the scheme. The Council element was a local authority-financed conference and exhibition hall, plus an underground car park. A privately-financed element was for shops, offices and an hotel. Following the withdrawal of the private developer, the Council assumed financial responsibility for the whole complex.

The major landmarks in the project are summarised in Table A.1. Some of its features are:

(a) In terms of duration, this is not a simple project. From initial market survey to completion exceeds 10 years, with a construction period of some 5 years. Such a timespan puts the centre in the same 'league' as the development period for advanced technology aerospace projects. Delays have also been substantial, with a time slippage factor exceeding 2·0.

(b) There have been major changes in the project, both before and during construction.

(c) Claims about the reliability of the cost estimates are fascinating. In 1978, it was officially claimed that 'estimates will become more reliable as the contract proceeds towards completion'. In July 1979, councillors were reported as claiming that the final price 'will not reach £17 million' and such an estimate 'was entirely dreamed-up'; by March 1980, the latest estimate was £18·25 million!

(d) The centre provides an opportunity to compare council and private decisions *on the same site*. Before the Council assumed total financial responsibility, private markets were willing to revise *downwards* their proposals and ultimately withdraw from the scheme. The Council started work on the site *before* private finance had been obtained ('it was important to make an early start'). Once started, it was difficult to stop and revise downwards: indeed, once the foundations for both the hotel and office were completed, the Council decided to continue with the offices even though it had no private developer. In other words, private market signals were available, but the Council chose to ignore them.

(e) Originally, the scheme was regarded as worthwhile on the basis of its 'substantial' social benefits (jobs, trade). In 1980, the District Auditor's Report on the scheme concluded that

> 'The benefits have never been evaluated in full and whether there is any satisfactory method of doing so must be open to some doubt'.[22]

[22] District Auditor's Report, *Harrogate Advertiser*, 29 March 1980, p. 9, para. 19.

TABLE A.I

THE PROJECT HISTORY

Major features	*Dates*	*Cost estimate £m.*	*Estimated completion date*
1. Consultant's study of *Future of Harrogate as a Conference and Exhibition Centre:*			
(a) Start of market survey	Nov. 1970		
(b) Submission of Report: the case for a new conference centre	June 1971		
2. Negotiations with private developers; first cost estimates; site selection and planning applications; negotiations with architects; design stage drawings	Sept. 1972 to July 1974	(1·085)	1975
3. Council approval of Maxwell Proposals I: Council-financed conference and exhibition centre and underground parking; a privately-financed commercial element of hotel, offices, shops	July 1974	(7·4)	
4. Competition started	April 1975		
5. Contractor selected	Feb. 1976		
6. Council approval of revised scheme: Maxwell II, no hotel and smaller office block	April 1976	7·81	Nov. 1978
7. Construction started	Aug. 1976		Feb. 1979
8. Modifications and revised estimates:			
(a) Revision 1	Feb. 1977	7·95	
(b) Private developer withdraws: Council proceeds with the whole scheme (Revision 2)	July 1977	9·65	
(c) Revision 3	Oct. 1977	9·95	
(d) Major upgrading of facilities (Revision 4)	April 1978	10·95	end-1979
(e) Revision 5	Oct. 1978	11·42	mid-1980
(f) Revision 6	July 1979	14·75	Sept. 1980
(g) Revision 7	Mar. 1980	18·25	May 1981

[*Notes and Sources are on p. 120*]

Notes to Table A.1:

(1) All cost estimates refer to the estimated cost of the whole scheme, namely the council and commercial elements.

(2) The cost figures in brackets are for earlier schemes and are in constant prices at the date of the estimate. All remaining costs are for prices to completion and include an allowance for estimated inflation.

Sources: District Auditor's Report, *Harrogate Advertiser,* 29 March 1980; 'Getting this complex matter in perspective', *Harrogate Advertiser,* 17 February 1979.

APPENDIX 2

A Model of Cost Escalation

Figure 1 outlines a model of cost escalation based on a trade-off between time and cost. Consider a project of a given quality or performance Q_0, estimated to cost C_2 and requiring T_2 years to complete. Actual costs can exceed estimates if the project is required faster at, say, T_1. Or, unexpected modifications might result in Q_1 being purchased at C_4 and T_4. Alternatively, optimism might lead a contractor to under-estimate costs (say, C_1 instead of C_2 for T_2); or, if Q_0 is uncertain, there are incentives to submit minimum or optimistic cost estimates.

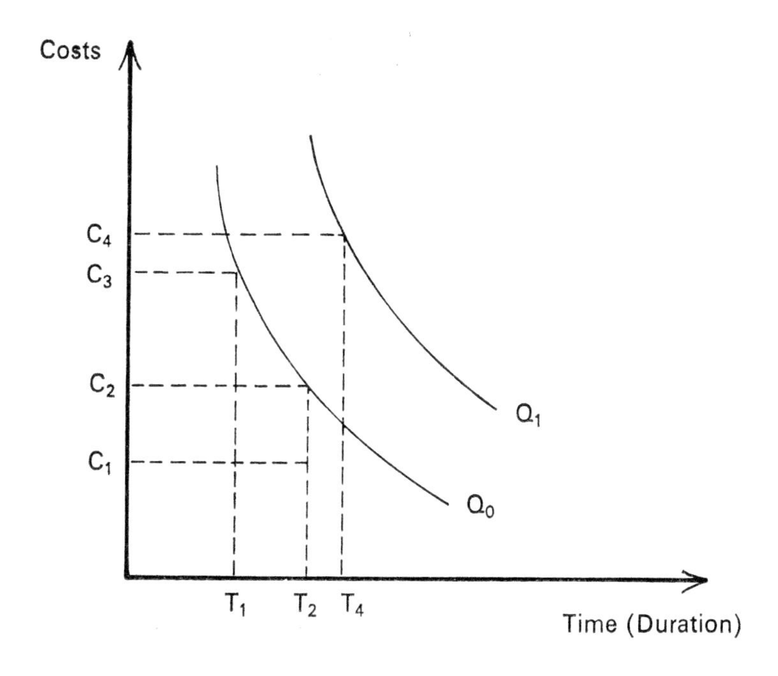

Figure 1: Projected cost escalation

The trade-offs in Figure 1 are *planning* schedules. Once a contract has been awarded, firms might behave as though the actual time-cost relationship is *positive*, which will be reflected in both cost and time slippages, with extra time being costlier. It might also be relevant to report some limited efforts at testing for the effects of the number of bidders on local authority contract prices. There was no evidence that, with the existing selective system, the number of bidders had any statistically significant *negative* effect on the minimum-priced tender.[23]

---

[23] $$P = -245{\cdot}8 + \underset{(33{\cdot}0)}{30{\cdot}6B} + \underset{(5{\cdot}8)}{\overset{**}{26{\cdot}8D}}$$

$$\bar{R}^2 = 0{\cdot}53$$

where P = minimum bid from the set of bids for each project. Units are £000's, the number of projects was 20 with an average of 5-6 bidders per contract. The data were from the early 1970s. Alternative price variables were used (e.g. price dispersion), with similar results.

B = number of bidders

D = actual duration of project (months).

Figures in brackets are standard errors; ** significant at 1 per cent level. (Further analysis and discussion is in K. Hartley and J. Cubitt, 'Cost Escalation in the UK', in Eleventh Report from the Expenditure Committee, *The Civil Service*, HMSO, London, HC 535, Vol. III, Appendix 44, 1977.)

Arthur Seldon (*Chairman*): *Merrie Cave knows quite a lot about public affairs; she also teaches history; and she wanted to get onto her local council but failed. Perhaps she will now tell us why.*

COMMENTARY

MERRIE CAVE

Standing as a Conservative candidate in the Holloway ward is like standing as a Communist in Belgravia. My interest in bureaucracy stems wholly from personal experience: living in the Soviet Republic of Islington; my father's battles to resist the expansion of his local government bureaucracy in the closing years of his career; and my daily work as a teacher for the Inner London Education Authority, one of the most Byzantine of all bureaucracies. I would like to make one or two general observations before commenting on Dr Hartley's paper.

Our public services today spring not only from socialist thinking, but also from the transfer of the welfare state from the Empire overseas to the Empire at home. For the last thirty years or so, affairs have been discreetly managed so as not to offend the natives, at the expense of efficiency and excellence. Countless people have actually abandoned doing any good for the natives at all in favour of protecting their own jobs—at the natives' expense and assisted, of course, by the union Moguls.

I am not very hopeful about Dr Hartley's suggestion that competition in contracts will keep prices down: our borough architect has openly admitted that there is a 'building ring' in Islington which keeps the price of rehabilitation up. The only solution in my view is to stop local authorities building altogether.

Bureaucrats' resistance to cuts

Most bureaucrats will resist to the last typewriter any attempt to shrink their empires. The Editor of the magazine *NOW!* recently quoted the instructions issued by the Head of the Civil Service to the mandarins answering questions from Select Committees. Some in Whitehall, it would seem, would like to run the country without the irritating participation of elected representatives. At town-hall level NALGO (the local government workers' union), in some boroughs, would like parity with councillors. Local bureaucrats are extremely adept at dodging 'cuts', as Dr Hartley has pointed out. Islington Council has just cut its £15,000 a year grant to Holloway Housing Aid—which provides advice. But the Housing Department is now providing extra counter staff to give the equivalent advice; the budget for that costs £18,000 a year.

Local government is now like a moribund animal, over-run with parasites. The only cure will probably be politically unacceptable. Years ago, public servants were relatively poorly paid, but provided an efficient service in exchange for a high (almost total) degree of job security. Now they have both job security and high pay. They enjoyed esteem from the public, whereas now they often face denigration and derision. If we seriously want to reduce the number of public servants, we must pay them less than they could earn in private industry or commerce: simple supply and demand. I am therefore deeply dismayed that the present administration seems to be perpetuating unjustified awards in the public sector and maintaining ludicrous commissions like Clegg.[1] Parkinson was right—it has all been said before; I cannot say anything else.

Questions and Discussion

EDWARD DENISON (*Leader, North Yorkshire County Council; solicitor; director of a number of manufacturing companies*): I wonder if I might take a couple of minutes to deal not only with matters arising from Dr Hartley's paper but also with some of the comments which have been passed today, especially of course as North Yorkshire has been named once or twice. May I also point out that Harrogate and its conference centre have, of course, nothing at all to do with the county council.

I was somewhat surprised, despite its title, that as the day went on the Seminar appeared to me to be developing into a swingeing attack on local government itself. A number of phrases were used such as 'We

[1] [The Prime Minister, Mrs Margaret Thatcher, announced in the House of Commons on 4 August 1980 that the Commission on Pay Comparability would be wound up 'by the turn of the year'.—ED.]

must do something'. Now who is going to do it? There are certain mandatory requirements, as Sir John Grugeon has already pointed out, which are incumbent upon local authorities. Mr Marsland, for example, said that real choice was being denied to real people. But that is brittle and shallow. Who are the real people? It is my submission that the real people are those who get out of the ivory towers of the universities and who are prepared to 'do something' in local government itself. How many people here are actively engaged in giving up their time to local government?

Accountability and bureaucracy—local or central?

We have heard a lot about accountability and why local government is not accountable. Accountability is a subjective term. It depends on how you choose to interpret it. And there was talk about local government and the alternative of central government. Central government itself, through Edward Du Cann and his committee, is only now getting round to talking about accountability. So what would people prefer? Would they prefer to be able to try and get action from their local councillors, whatever their calibre, or would they wish to see the whole thing go to Whitehall? There has been a lot of talk about bureaucrats. Again, how do you define bureaucrats? Are they university lecturers or are they all county planning officers? Or are they in fact people in Whitehall. If you do away with local government, then in practice you are going to find yourself in Whitehall. Are you going to be any better off with Whitehall bureaucrats and accountability than you are with local government?

Dr Hartley talked about cost escalation. I would ask him when he was last involved in the construction industry. If you insist upon fixed price tendering, ultimately it becomes more expensive than fluctuating tendering because people load a fixed price tender to take account of all foreseeable costs in the future. And with inflation running at the present rate, people are loading fixed price tenders to an even greater degree than before.

One other small point of Dr Hartley's—about competition. Of course there are aspects which need looking into, aspects which can be bettered, aspects we want nothing to do with in local government, aspects we would love to abandon altogether if the legislation would allow us to do so. But let us also be clear that competition does not automatically ensure that the service is going to be cheaper or more cost effective. As an example, let me suggest trying to offer the winter road maintenance of some of the North Yorkshire moors to the private sector. Either the service will not be forthcoming, or, if it *is*, it will be as a loss leader to hook you for an increased price the following year. Of course, one must keep testing what we do in local government by reference to private sector alternatives, but the private sector is not an automatic panacea for all the ills of local government.

Finally, coming back to my opening theme, what it really boils down to is the calibre of person that you can attract into local government. If your alternative is central government, are you really happy with the calibre of person that is sitting on the benches in Westminster? I am not happy with either, but that is what it comes to; you have got to make do with what you have got. One accepts a lot of the criticism that has been pushed out today, but if you are going to criticise you ought to offer constructive alternatives. And I do not think the alternatives are any better than what we have at present.

JOHN TYNDALL: I am a ratepayer and a resident of south Oxfordshire' a lovely rural county. We have a Tory-led council. But our local district council is not under control in any sense in which a businessman might understand the word. In business you have the most rigorous examination of any capital expenditure project, and a proper economic analysis of the information. You also have, of course, a most careful analysis year-by-year of your manning levels. It was a shock and a surprise to find that a Tory-led council in south Oxfordshire is not under control in these respects.

ARTHUR SELDON: I suppose we cannot avoid the use of party labels but we are talking about the principles of government rather than party philosophies.

BRIAN MICKLETHWAIT: I would like to reply to Mr Denison about those people in the wilds of Yorkshire. The free market pressure to which they would be subjected if the council did not mend their roads is a good thing because it might encourage them to live somewhere less costly to the rest of society. One of the great virtues of the free market is that it makes people think about the resources they consume. The council protects them from such decisions.

RICHARD LAMB (*Journalist; ex-county councillor*): The social workers have taken a terrific bashing today, probably with good reason, but I can remember the Curtis Committee Report, in the distant days of the Attlee Government, which showed that the way in which we looked after children in this country was appalling, and that the voluntary organisations simply could not cope with the problem despite the efforts of the NSPCC, the Salvation Army, the Church of England, Roman Catholic rescue societies, and so on. Children were having a very bad crack of the whip. The Curtis Report showed, for example, that when a mother or a father was imprisoned, quite often even in progressive authorities the children were simply kept in hospital or workhouse beds all day because there was no-one to do anything about it. That could go on for weeks or months. I remember when Children's Officers were first appointed in those dim and distant days when the welfare state started, and no doubt they have done an amazing amount of good, although there are far too many of them now. Do some people really contend that we can leave it all to voluntary enterprise? If so, I most heartily disagree. Does anyone think that we can get rid of local authorities altogether and have their functions carried out by central government? I disagree with that, too. Local government has done a very good job, but it has overreached itself now. Surely what we are talking about today is reform, not abolition, and not returning local government to private enterprise or to voluntary services.

P. BUTLER (*DoE*)**:** On the question of central *versus* local autonomy, one point has not had an airing today. There has been some discussion about the bad side of conformity. Perhaps there is a good side to it. I am thinking here of certain regulatory aspects of activities that are currently the responsibility of local authorities, at least in their implementation: planning matters, building regulations, and so on. Are we really saying that we want widespread discretion across the country on these matters? If so, we should be clear about the implications.

KEITH HARTLEY: Let me deal first of all with definitions. I did define bureaucracy in my paper as non-profit organisations in the public sector, including state agencies such as the Manpower Services Commission. Mr Denison was worried about my experience of fixed-price tendering, but experience of competition and fixed-price contracts tends to refute his argument. This is an area which raises fundamental and worrying problems, especially in defence and high technology, and I am not certain that the problems in local government are so big. As for the worry that fixed-price tendering would lead to increased costs through loading, my reply would be: 'yes', unless there was competition as a policing mechanism. With competition, comparative policing should give us a more accurate indication of the true cost of a project. If it is going to be expensive, there is no way it can be avoided—it will be expensive. I would rather have that information conveyed to councillors at the very outset. Indeed, if one of the implications of a firm fixed-price contract system is that a project looks very expensive, well, that's the name of the game; and if it means the project does not go ahead, so be it. That is one of the benefits behind a firm fixed-price contract system with competitive policing. My preference is for open competition to avoid the problems of collusion.

Two final points. I really do have a worry about how we can assess efficiency in the public sector, despite Mr Denison's reservations about the alternative I was proposing. Is there any means we can use to assess efficiency other than the market-based solutions I formulated and other people discussed? What is Mr Denison's preferred method for assessing efficiency?

Mr Butler made the point about regulation. The difficulty here is that some of the economic explanations of regulation in more recent models suggest that it can give perverse outcomes, often favouring producers rather than consumers. Finally, one of the policy guidelines I proposed towards the end of my paper amounted to asking: What is the case for local government; what might be its minimum functions? I have not heard people say: 'Well, perhaps it's police, street lighting, civil defence'. Should we go beyond those functions and, if so, where do we stop?

PART VI
IS LOCAL GOVERNMENT WORTH SAVING?

8. How To Save Local Government

G. W. JONES

London School of Economics and Political Science

The Author

GEORGE W. JONES: Professor of Government, London School of Economics and Political Science. Formerly Lecturer in Government, University of Leeds, 1963-66. Member of the Layfield Committee on Local Government Finance, 1974-76. Author of *Borough Politics* (1969). Co-author, *Herbert Morrison: Portrait of a Politician* (1973). Joint editor, *Political Leadership in Local Authorities* (1978); editor, *New Approaches to the Study of Central-Local Government Relationships* (1980).

Commentary Author

ANTHONY BAYS: Member of Lambeth Borough Council since 1974. Educated at Oakham School, Rutland, and the University of Bradford, where he studied economics. Investment analyst with stockbrokers Spencer Thornton since 1976.

I. LOCAL GOVERNMENT: 'AN ESSENTIAL ELEMENT OF A PLURALIST SOCIETY'

Why is local government worth saving? Because it is an essential element of a pluralist society, a counterweight to a monocentric government, and a check on centralisation and imposed uniformity and standardisation. It is a means through which citizens can influence and control the decisions that affect them collectively; it helps to distribute influence in society more widely and to challenge power wherever it accumulates to ensure it is exercised responsibly and responsively.

Dangers of over-centralised power

Too much power can be concentrated in central government, as it can in central private organisations, so that it may act irresponsibly and unresponsively. The consequences may be:

(a) to damage the centre, overburdening it with activities that distract it from its proper national and strategic rôle;

(b) its services may not suit local needs and conditions, and may not reflect local wishes;

(c) individuals may become frustrated and alienated as they recognise that decisions affecting them are taken by remote bodies scarcely accountable to them;

(d) the more decisions are concentrated at one central point, the fewer the opportunities for public participation in government, for involvement in and for influencing the decisions that affect citizens collectively;

(e) social advance is impeded by the laying down of national standards and the imposition of the single correct line. Local creativity, experiment and pioneering to solve social problems are obstructed by the omniscient centre. Social learning is blocked;

(f) resources may be wasted if artificial standards are applied without variation to local need;

(g) the perpetration and perpetuation of the one big error are encouraged.

Such dangers may be checked by pluralistic decentralisation of power, which can set up countervailing pressures to the centre, not to overwhelm or abolish it, but to provide balancing forces. The aim is not to let loose rampant and uncontrolled localism, and the sharpening of geographical inequalities, but to provide a more balanced array of pressures in public policy-making.

Centralised economic control is counter-productive

Centralisation—the strong central state—has been justified as essential for co-ordinating the management of the economy (national economic planning), for achieving functional (service) efficiency, and for ensuring an equitable distribution of services. In theory these objectives are important, but in practice centralisation has now gone so far as to prevent their achievement. Centralisation has become counter-productive.

In trying to control the economy the centre has sought ever more detailed controls over local government expenditure, penetrating into individual local authorities' decisions on service provision, capital projects, spending, staffing and rating. For managing the economy the government does not need to control local government current expenditure in detail or to determine broad totals of local government current expenditure. *It should set limits to local authority borrowing and fix the level of central grant, and then allow local authorities to spend what they like, as long as it is financed through local taxation that bears on local voters.*

The proposition italicised above is not part of conventional wisdom. In discussion about central-local relationships there is general agreement between central and local government that the centre needs to control the total of local government current expenditure. For macro-economic purposes, it is argued, the centre must determine the aggregate level of local spending. This orthodox view, accepted on Treasury advice by the Layfield Committee[1] and by the Chartered Institute of Public Finance and Accountancy's (CIPFA) report,[2] forms the basis of Treasury concern for local government

[1] Cmnd. 6453, 1976, p. 239, para. 7.

[2] *Local Government Finance and Macro-Economic Policy,* CIPFA, 1977, pp. 10-12.

expenditure, its resistance to a local income tax, and is a justification for many detailed central controls over local government.

This conventional wisdom ought to be challenged. It is hazardous for a non-economist to trespass on the preserves and jargon of economists, but sometimes the non-specialist may notice aspects missed by specialists.[3] At least he may raise questions for defenders of current orthodoxy to reply to and justify their case for strong central controls over local government expenditure. Indeed, they seem to regard expenditure by local authorities as no different from spending by government departments, treating local government as an outpost of central government, despite the constitutional, political and electoral separation of the two levels of government.

II. MACRO-ECONOMIC STABILISATION POLICY AND LOCAL GOVERNMENT

What is meant by macro-economic policy? A useful description has recently been provided by the Ball Committee on Policy Optimisation.[4] It is concerned with fluctuations in the level of economic activity, involving control over the level of employment, the rate of inflation and regulation of the balance of payments. It is often called 'stabilisation policy'.

Orthodox arguments for central control of total local spending

Three reasons are commonly given why the Treasury for macro-economic purposes needs to control the total of local government current spending:

(a) Local authority spending affects the level of aggregate demand in the economy, and hence employment, inflation and the balance of payments.

(b) Local authority spending, insofar as it is financed by rates, affects the cost of living and hence will influence wage claims.

[3] I must acknowledge stimulating conversations with R. A. Jackman of LSE and Professor P. M. Jackson, University of Leicester. (cf. R. A. Jackman, 'Monetarism and Control of Local Spending', *C.E.S. Review,* September 1979, pp. 29-31.)

[4] *Policy Optimisation Report,* Cmnd. 7148, 1978, pp. 14-15.

(c) Central government, wishing to limit rate increases, will be obliged itself to finance increased local government spending by higher grant. Thus the centre must control local government spending in order to control its own expenditure.

(a) Does local authority spending affect macro-aggregates?

If extra local spending is financed by higher local rates, the consequences are that there will be less private consumption, less business expenditure or less government spending. A re-allocation within the total will have occurred, not an increase in total expenditure. The orthodox reply is that, while high rates may reduce other components of demand, there is no guarantee that the fall in other components will precisely equal any increase in local government spending. Part of the rate bill may be paid from household or business saving rather than from reductions in spending. It is this line of argument which leads to the theorem of the 'balanced-budget multiplier', which states that 'equal increases in spending and taxation increase demand, while equal reductions in spending and taxation reduce demand'.

But macro-economic policy is not a precise science. There are uncertainties both in economic forecasts and in estimates of the impact of any policy change. As CIPFA has shown,[5] local government expenditure is the smallest of the main components of aggregate demand in the economy. Even a quite significant change in local government spending which is mostly—even if not necessarily entirely—offset by the impact of higher rates can therefore have only a very small impact on total demand, certainly well within the range of forecasting uncertainty.

So, why is the Treasury so concerned about controlling local government expenditure more than the more substantial elements? Further, even Treasury officials have stated that there is a wide margin within which government expenditure can vary without being de-stabilising, e.g. ±£1,000 million at 1975 prices.[6]

As long as local government finances the extra spending out of its own tax, and balances its budget without resort to borrowing to cover deficits, its expenditure poses no problems for the public sector

---

[5] *Op. cit.*, p. 10.

[6] Fourteenth Report from the Expenditure Committee, *The Motor Vehicle Industry*, HC (1974-5) 617—II, pp. 201-2.

TABLE I

BRITISH ECONOMY: MAIN COMPONENTS OF AGGREGATE DEMAND, 1976

	%
Consumers' expenditure	58
Central government purchases of goods and services	15
Local government purchases of goods and services	13
Investment (other than by central and local government)	16
Exports	31
–Imports	–33
	100

Source: CIPFA, *op. cit.*, p. 10.

borrowing requirement, monetary management, imports, exports, or sterling, and the aggregate level of demand stays the same. Monetarists will observe that local authorities' current expenditure in these conditions cannot influence the money supply.

(b) **Does local authority spending affect the cost of living?**

It is surely inappropriate to include rates in the retail price index. As a tax they are like income tax, which is not included. Some argue, however, that they are more like VAT, a consumption tax, which is included. The consistent approach would be to include all taxes or none, and not to give an advantage to central government by excluding income tax. The case for taking rates out of the retail price index is strong. The point of the index is to measure changes in the cost of purchasing a given set of goods and services. If rates rise because of an increase in real local government spending, it is not the price of the given set that has risen; rather the number of items in the set has increased.

In any case, since domestic rates on average hover around only 2-2·5 per cent of household disposable income, even large rate increases would not add significantly to the cost of living. If the prevailing rate of inflation is around 10 per cent, even a rate increase on average of 20-25 per cent will add only about one-quarter of 1 per cent to the retail price index. And, even now, with variations in rate increases around the country, there is no correlation between higher rate increases and demands for higher wages. Therefore (b) like (a) is exaggerated. Monetarists will observe that inflation and wage growth depend on the money supply.

(c) Must government control local expenditure to control its own spending?

There is no need, for macro-economic purposes, for the centre to step in with higher grants to finance increased local government expenditure. It is more appropriate for it to control its own (grant) expenditure, than to seek to control local spending over which it has no direct authority. It is strange that the government's annual white paper on public expenditure, based on the PESC process,[7] includes total local government expenditure for four years ahead but not the one item that central government can control, its grant to local government.

My proposition is that, as long as extra local government spending is financed out of its own tax, and it balances its budget without recourse to borrowing or is not bailed out by central grants, then this extra local expenditure need have no effect on aggregate demand or the money supply. Hence there is no need for the Treasury to try to control aggregate local expenditure in pursuit of macro-economic policy.

Government may, however, have other objectives which push it into intervention and cause it to treat the expenditure of local authorities as if it were no different from the expenditure of central departments. Some argue that it is impossible to disentangle the management of the economy from these other concerns of government, including the use and distribution of resources, regional policies, industrial sector strategies, service priorities, as well as its political aspirations. These may all be interdependent and are jumbled up in policy-making, but they are conceptually distinctive, and there may be important advantages to be gained from distinguishing macro-economic management factors from the rest. Policy-making will be less confused and ambiguous: both central and local government will be clearer about the reasons for certain policies. And it will be more difficult for central government to seek to achieve other

[7] PESC stands for Public Expenditure Survey Committee, a group of senior civil servants who co-ordinate departmental spending plans.

objectives under the guise of its need to control the economy for macro-economic purposes.

And, as Layfield noted (Appendix 5), some other European countries, often with more healthy economies than ours, do not practise such tight central controls over local government expenditure. Does Britain really need to?

Local experiment and learning frustrated

In trying to achieve functional (service) efficiency the centre has again become counter-productive, hindering local experiment and social learning. The setting of national standards, even of minimum standards, especially of an 'input' nature, prevents creative thought about how to tackle social problems.[8] What is required, particularly in a period of resource squeeze and growing social dislocation, is for *local authorities to experiment in devising performance criteria and 'output' standards, and for the centre to aid and facilitate the dissemination of the results.* The latter should not seek to impose its view of what should be done.

The attainment of more equality through redistribution is a central government task, but again the centre's exertions have been counter-productive. The rate support grant industry (and even more so the future block grant industry) grind on in baffling complexity, distorted by damping, claw-back, safety nets, multipliers and partisan fiddling. Because central grant looms so large the centre is encouraged to intervene in local government, and it seems crucial to produce an accurate mode of calculation and distribution.[9] A large amount of time and effort is spent on getting grant precisely correct (which it can never be), with the consequence that changes occur each year. The resulting instability weakens local responsibility, since changes in the rates levied by local authorities bear little relationship to changes in their spending decisions, and are more determined by changes in their grant.

The objective of the centre should be to put each local authority on the same level, in terms of resources and spending need. Grant should be designed with that one objective in mind, and should not be

[8] A critique of 'standards' is to be found in G. W. Jones, 'Central-Local Government Relations: Grants, Local Responsibility and Minimum Standards', in David Butler and A. H. Halsey, *Policy and Politics*, Macmillan, 1978, pp. 74-78.

[9] Jones, *ibid.*, pp. 70-74.

burdened with other central aims. It should not be used as a general subsidy. If this change is made, then grant can be reduced from around 60 to around 40 per cent, and perhaps even further.

III. MORE LOCAL TAXES FOR MORE LOCAL DEMOCRACY

Local authorities should raise more of their income from their local voters, so that a closer link can be forged between pressures and decisions locally to spend more or less *and* pressures and decisions to have more or less local taxation. A better balance between these pressures should attain more responsible decision-making than at present.

Three proposals for responsible local decision-making

With this principle in mind three proposals are essential:

(i) *The non-domestic rate should be transformed into a national tax.* It is not appropriate for a local tax, since it does not bear directly on local voters.[10] Its transference to the centre would provide the national government with a useful instrument for its macro-economic, industrial and regional policies, and could be linked with government policies for employers' national insurance contributions and VAT.

(ii) *Domestic rates should be retained as a local tax, and should be developed as an element of a wealth tax, based on capital not rental valuation, and perhaps renamed to emphasise this aspect.* It is an ideal local tax, bearing on local voters.

(iii) To make up for the revenue lost by the reduction of grant from 60 to 40 per cent, *a local income tax should be introduced, with the rate of tax determined by the local authority.* It, too, is an ideal tax for local government, since, if based on place of residence, it will bear on local voters. There are at present about

[10] A critique of non-domestic rating as a local tax is in G. W. Jones, 'Some Post-Layfield Reflections about Non-Domestic Rating on Commercial and Industrial Properties', in R. A. Jackman (ed.), *The Impact of Rates on Industry and Commerce*, C.E.S. Policy Series, No. 5, 1978, pp. 37-50.

15 million domestic ratepayers and 24 million income tax payers. Thus 9 million people are contributing to the finance of local government through their national income tax, fed into grant, but they do not realise they are doing so. A local income tax (LIT) would make them aware of their contribution to local government and reduce the complaints about earning non-householders who are so often alleged not to be financing local government.[11]

This method of financing local government, through reduced grant, domestic rates and local income tax, could be fitted into the present structure of local government. Local income tax would be allocated to the big-spending authorities, metropolitan districts, outer London boroughs, shire counties and Scottish regions; while domestic rates would be sufficient for the low-spenders—shire districts and metropolitan counties. Inner London, as always, presents a problem, in the absence of a body like the LCC, but perhaps London government cannot avoid a further reorganisation.

If the structure of local government generally were reformed into regional and town governments, or into unitary governments, then the allocation of the two local taxes could be achieved more easily and cheaply. Indeed one of the criteria for reform of boundaries and functions should be the need to produce units that can be based on financial arrangements that promote responsible local decision-making.

The role of direct charging

Direct charging for services is another way of providing some income for local government, and one that might ensure that functions are responsive directly to consumers. Sceptics have tended to be hostile to this method, because it seemed to give the advantage to those with the biggest purses. If national government ensures a more equitable distribution of income and wealth, however, charging for certain services becomes more attractive.

The Layfield Committee received evidence from Ralph Harris and

[11] The cost of implementing LIT could be reduced below the Layfield estimate through self-assessment: N. A. Barr, S. R. James and A. R. Prest, *Self-Assessment for Income Tax*, Heinemann, 1977, pp. 195-6.

Arthur Seldon about the rôle of charging[12] and concluded that there could well be scope for financing a bigger share of local government expenditure from charges. It felt, however, that the extent and level of charging and the balance of policy considerations lay outside its terms of reference. It called for a review of the policies of both central government and local authorities towards charging for local services. In 1977 the Consultative Council on Local Government Finance set up a Joint Working Group on Fees and Charges, composed of both central and local officials. It was excluded from examining housing rents, transport fares and school meals and was limited to existing fees and charges. Its report, which contained a checklist of questions on the determination of charges, was never published. My judgement is that the issue of charging raises such fundamental political, social, administrative and economic questions that a Royal Commission is required to examine 'charging' in central and local government.[13]

IV. CONCLUSION

The purpose of this paper is to outline a set of proposals to sustain a responsible and responsive system of local self-government. It concentrates on financial aspects, particularly those usually monopolised by economists. It concentrates most on the issue of what controls the central government needs over local government expenditure and its financing to achieve proper national objectives. It adopts the approach of the majority of the Layfield Committee that there should be a shift of responsibility away from the centre and more to local government. It also takes an approach completely opposed to the Local Government, Planning and Land Bill being promoted by the present Conservative Secretary of State for the Environment, Mr Michael Heseltine. It appears that he is intending to increase centralisation so that the central government will be

---

[12] [Reproduced in Harris and Seldon, *Pricing or Taxing?*, Hobart Paper 71, IEA, 1976.—ED.]

[13] Layfield Commission *Report*, Ch. 9, pp. 132-42; Arthur Seldon, *Charge*, Temple Smith, 1977; Consultative Council on Local Government Finance, Report of the Joint Working Group on Fees and Charges, CCLGF(79)2, 1979; and J. G. Gibson (ed.), *Fees and Charges in the Personal Social Services*, Institute of Local Government Studies, Birmingham, 1979.

deeply involved in the services, expenditure and rating decisions of each local authority.

But he is also pursuing a policy at variance with the dominant monetarist line of his Government. As I have suggested (pp. 134-136), a monetarist economic stance does not require the tight controls over local authority expenditure as embodied in the present Bill. Should not Sir Geoffrey Howe and Sir Keith Joseph take Mr Heseltine on one side and explain that his Bill offends against the basic economic principles of the Government as well as angering and alienating usually loyal Conservative supporters in local government?

Arthur Seldon: *Now a councillor from a council with a very interesting record—Lambeth.*

COMMENTARY

ANTHONY BAYS

I would like to comment on what I have seen in local government in a somewhat different type of authority from those we have heard about this afternoon. Arthur Seldon made a very significant point when he said that the differences are really between those of us who do and those of us who do *not* want government to do things, rather than between those of us who argue about the different *ways* in which government can do things. The problems presented in an inner urban authority, such as the one to which I belong, Lambeth, and a mixed authority such as Kent County Council, where there must be industrial towns as well as large residential areas, and perhaps North Yorkshire, which I would imagine is a more rural authority, are different. These differing authorities would obviously look at the market and the functions it could carry out from among their local government responsibilities rather differently. And that is where I think the most important part of the decentralisation argument should come in. I am concerned that we are still talking about whether meetings should take place in Marsham Street or in Whitehall, rather than saying: 'Should those meetings take place at all?', or 'How many subjects should those meetings cover?'

Limiting local government—and the interest-groups

Community involvement, neighbourhood councils, Crosland—all that sort of thing—produced a great era of local authority growth, and it is still there. I think we have got to say we are going to examine the problem of decentralisation. It is as much a question of limiting the functions of local authorities—saying that this is where we have to turn the clock back—as of shifting the burden away from the industrial rate-payer towards a local income tax. In relation to my theme of decentralisation, I would also like to note that we seem to be getting more of an interest-group type of confrontation in our local politics, as well as in our national politics. It is a major factor facing the Government in its management of the public sector. The trend which has been developing is for everyone to make sure he is in an interest-group which can fight its way. Few people now question whether belonging to a strong organisation is the best way to advance their own interests—rather than, for example, by establishing their own business. Thus there is an aspect of the decentralisation argument which is different from those discussed so far.

Having seen the turn-out at some of the Ratepayers' Association's meetings following the 30/40/50 per cent domestic rate increases in inner London boroughs this year, I am sure there is a lot of logic in Professor Jones's argument that the domestic rate is extremely low. But I think it always will be a very unpopular tax, and I can see there are arguments in favour of moving over towards a local income tax system. But I think we have reached a point where the interest-groups are preoccupied with fighting each other over the existing system. I am not sure how interested they are going to be in the radical proposal which Professor Jones has made.

But my main conclusion is that we have really got to start thinking in terms of how to curtail the areas of local government involvement as much as simply changing the details of its present organisation.

Questions and Discussion

PETER BROOKS (*Thomas Tilling*): First, does the low turn-out at local government elections imply that people are not interested in local government, or perceive that local government is not very important, or do not want local government? Secondly, do you believe that changes in local government, in particular more accountability and/or higher levels of local taxation or finance, would increase electors' interest?

PROF. JONES: The introduction of local income tax would make more people aware that they are making a major contribution to the financing of local government and this would be an incentive for them to turn out and vote. Currently, some 9 million people are helping to finance local government who are not aware of it. A local income tax would make some of them turn out to vote.

As to why people vote, volumes have been written on the subject, and there is even a so-called discipline of psephology that studies it. I think one of the reasons for the low turn-out is that many people are not aware they are contributing to local government. They realise they are getting services from local government and on the whole are satisfied. That seems to be the evidence of public opinion polls. The overwhelming impression from polls is that ordinary people are satisfied with council services, though to repeat, not all know they are helping to finance them.

JOHN FAWKNER: We have concentrated on financial structures, but could it be that changes are necessary in the process of democratic election for local authorities? At present the system is highly indirect; you vote for a local representative who then helps in the choice of the chairman of the education committee and so on. Might it not be better, more direct, involve more people and imply much better accountability if people actually voted for candidates for the office of chairman of the education committee? It will be countered that the individual voter will not know enough about the candidates. I would answer that, by and large, in elections people vote for rejecting what is past rather than for what they expect to be new. I think this idea merits serious consideration.

GEORGE JONES: I would agree about the need to study the electoral arrangements of local government. Personally I would have thought there is a very good case for introducing proportional representation into elections for local authorities. But I am not sure I agree with the view that the local electorate should be invited to choose chief education officers or the chairman of the education committee. Those sorts of decisions should be left to the specialists in politics, namely, the leading figures in the local council.

P. BUTLER (*DoE*): You do not need a grant from central government to implement equalisation. You could merely distribute it between authorities. Therefore the figure of 60 to 40 per cent is not particularly relevant. Professor Jones rather ducked the issue on equalisation, given there is open-ended local authority current expenditure but close-ended central government grants. Unless you have close-ended grants to individual authorities, which, of course, is a primary objective under the unitary grant system in the current Bill, the process of setting rates will have the effect of redistributing grants. Those of us familiar with the old Rate Support Grant process will know about 'clawback' under the old resources elements and, therefore, there is an equity element. How do you reconcile this point about non-interference with individual local authorities but nevertheless ensure equity between individual authorities? During the 1970s it was one of the local authority associations which protested most strongly about the open-ended nature of grant.

PROF. JONES: The figure of 40 per cent that has been queried emerged from the advice of the technical experts of the Department of the Environment who told the Layfield Committee that, under the present arrangements, a grant of only 40 per cent did the job of equalisation. The rest, then at about 21 per cent or so, was, as it were, a *per capita* handout—almost a broad general subsidy. They told us that the job of equalisation, putting all local authorities on a broadly similar level in terms of spending need and resources, could be done under the present arrangements with a grant of 40 per cent. In my paper, I said the grant could be reduced from around 60 per cent to around 40 per cent and perhaps even further.

On your second point: Under any system of redistribution there is implicit in the calculations a notion of what each authority ought to be spending. The major change that is proposed in the Bill is to make the calculations of what authorities should be spending explicit. That changes the entire nature of the process, because a Minister may have to stand up and say what a local authority ought to be spending, perhaps in response to a parliamentary question. Thus it will change the *locus* of responsibility. Although it is superficially a change of presentation, the fact that it is now being brought out into the open means a fundamental change of principle and a shift of responsibility for what a local authority ought to be spending from local to central government.

DAVID KING (*University of Stirling; member of CIPFA panel on macro-economic policy for local government finance*): I would dispute Professor Jones's thesis that the CIPFA panel argued the case for more central government control over local government expenditure. In our report (entitled *Local Government Finance and Macro-economic Policy*) we argued that central government did not need *control* but rather what Professor Jones was advocating, namely, *influence* over local expenditure. Moreover, in arguing for further influence over local expenditure by means of a completely fixed grant, he is not allowing the grants to vary with local tax rates at all, and he proposes completely fixed limits on local authority borrowing. Our report concluded there was no need for any further controls beyond the present ones. So rather than we being the controllers and Professor Jones the liberal, the position is actually the reverse.

There is also a problem with the closed level of grants he suggests, whereby each authority's grant would be fixed and independent of tax rates. That would mean that the *average* authority today would have to raise its rates by 3½ per cent to increase its rate- and grant-borne expenditure by 1 per cent. Many authorities would have to raise their rates by 6 or 7 per cent to increase their rate- and grant-borne expenditure by 1 per cent. That would make it very difficult for local authorities to alter their expenditure. It would also result in some inequality between areas, because some would find it harder to alter their expenditure than others. I am not, therefore, in favour of that type of grant. I also wonder why it is so important to control local authority borrowing to the nearest pound: surely, if they borrow in the market-place, there is no more need to control their borrowing than to control industry's borrowing?

PROF. JONES: CIPFA acceptance of the Treasury line is in pp. 10-12 of their report. Would it make any difference that in my scheme the grant would be much smaller than now because of local income tax and, therefore, through the gearing effect, there would not be the disastrous consequences for individual local authorities? My scheme is based on the assumption that grant is reduced and replaced by a comprehensible method of financing local government, namely, local income tax.

TYRELL BURGESS: From my point of view we have just reached the germane point of the day's proceedings. Professor Jones's address is absolutely crucial. I wholly agree that it is important for the health of our society that the financing of local authorities is visible and therefore encourages democracy and accountability. I do, however, think that there is something very urgent in front of us which has been mentioned by Professor Jones—the block grant. I agree with him that this represents a fundamental, constitutional shift and one that has far-reaching consequences beyond the technicalities of the grant. What is profoundly depressing in the contributions of the local authority people here today, including that of Sir John Grugeon, is that the local authorities have, by and large, collapsed in the face of this threat. The challenge to all of us is that each of us individually has to try to decide whether or not we still wish to belong to a society of free people, and to that end we must begin to work for the destruction of this grant as soon as it is there to destroy.

ALFRED SHERMAN: I wish to belong to a free people and therefore I think we should abolish the parasite state both in Whitehall and in local government. I used to accept much of Professor Jones's idealised picture of local government autonomy, until I began to learn something about local government. I found, first of all, that the idea of local government autonomy is based on a contradiction, a fallacy. So long as local authorities provide many important services, the political will of the country will be for them to provide a basic level in such things as education, housing and the rest. Local government will therefore be financed centrally because the poorer authorities generally have the most need and, for one reason or other, generate less income. Thus you have national government finance; and where you have national government finance you will have national government control.

Secondly, there is very little democracy in local government. With few exceptions, local authorities are self-perpetuating oligarchies, part of the parasite state which grows and grows at our expense. Good people will not enter local government, not because it does not pay them enough, but for two other reasons: firstly, because they find the way it is run wholly uncongenial; secondly, because the oligarchy will keep them out since the people of both parties who control local authorities are generally low-grade people and Gresham's law (the bad drives out the good) operates in local government. Furthermore, there is much crypto-professionalism thanks to the mistake of attendance allowances and expenses.[1]

Lastly, we are stuck with the two monsters of Whitehall and local government. I believe we can begin to cut down local government first by privatising nearly all of its services.

PROF. JONES: On the first point about equalisation, I have said that I am not trying to abolish grants totally: I see the rôle for the grant as being to achieve equalisation. We can achieve that with a grant of no more than 40 per cent and it could go down much further. I have no idealised view about local democracy and local government, nor have I an idealised view about central government and central democracy. I am not working in the ideal world. I am trying to devise pressures that will counterbalance each other and provide checks, particularly checks on local oligarchies. Local oligarchies and central oligarchies will both flourish if there is not an informed, educated electorate to check and challenge them. I deplore the attacks on local government because they distract attention from the main villain of the piece, which to my mind is central government. We should direct our attention to exposing what goes on in Whitehall behind closed doors. We know much more about local government because it is more local, more exposed, more open. The press can get at it and expose its abuses, but who can tell what goes on behind the closed doors of Whitehall?

I take your point about professionalisation. One of the major problems in local government is the professionalisation of the officers. The danger is that they will become so professionalised that they will get out of control. I would advocate professionalisation of councillors so that there is a full-time career for councillors. What this country lacks is a local élite of dedicated professional councillors, like Herbert Morrison. But, as we have heard, they land up in the poverty trap if they try to live on their allowances. If we had more professional local councillors, we might improve the calibre of our Members of Parliament and Ministers. In other countries with federal systems, the national politicians have often cut their teeth and learnt their trade first in local or provincial government. When later they become national federal politicians, they are much more skilled in the art of politics. The current hostility to political skills and to the rôle of the politician in this country is very depressing. One way to

---

[1] Alfred Sherman, *Socialism. The Newest Profession*, Aims, 1978.

improve it is to have paid elected councillors, so as to raise professional political control over the bureaucrats.

SELDON: On that note, that it is central government which is an obstacle to advance, perhaps we can all agree. And in thanking our speakers I would suggest you applaud yourselves for your questions, your comments, and your good humour. Thank you very much.

OLIVER STUTCHBURY: If I may, I would like to propose a vote of thanks to the IEA for organising such a seminar on this all-important subject, and in particular to you, Sir, for having chaired the meeting so economically and competently.

Statistical Appendix

TABLE A

UK EMPLOYMENT ANALYSED BY SECTOR, 1965 TO 1978

	1965	*1970*	*1975*	*1978*
		thousands		
Private sector	19,217	18,262	17,664	17,545
Public sector:				
Public corporations	2,028	2,027	2,033	2,061
Central Government:				
HM Forces	423	372	336	318
Civilians	1,370	1,533	1,906	1,991
Local authorities	2,154	2,559	2,993	3,013
Total employed labour force	25,192	24,753	24,932	24,928
Local authorities as % of total employed	*8·6*	*10·3*	*12·0*	*12·1*

Source: Compiled by IEA staff from *National Income and Expenditure* 'Blue Books'.

Note: Part-time workers are counted as 1.

TABLE B

LOCAL AUTHORITY WORKERS IN GREAT BRITAIN: BY SERVICE, MARCH 1980

	*Number**
Education:	
Lecturers and Teachers	638,500
Others	467,700
Construction	152,200
Transport	31,400
Social Services	236,900
Public Libraries and Museums	37,000
Recreation, Parks and Baths	88,900
Environmental Health	24,200
Refuse Collection and Disposal	59,900
Housing	55,400
Town and Country Planning	24,100
Fire Service	46,000
Miscellaneous Services	299,000
Police Service:	
Police—all ranks	127,300
Civilians	48,100
Probation, Magistrates' Courts, etc.	18,600
	2,355,200

*Rounded to nearest 100.

Source: Dept. of Employment, *Employment Gazette,* September 1980.

Note: The figures are for 'full-time equivalents' of full- and part-time workers.

List of Participants at IEA Seminar on 'Local Government or Instrument of Central Government?' *(4 June 1980)*

ACLAND, ANDREW, *Centre for Policy Studies*
ALLPORT, J. A., *St Paul's School*
ANDERSON, DIGBY, *University of Nottingham/Social Affairs Unit*
ANDERSON, JUDITH

BAILEY, M. L., *RTZ*
BARNES, ROY, *LA Management and Computer Committee*
BAYER, SIXTO
BAYS, ANTHONY
BEALE, NEVILLE, *Member of ILEA Education Committee*
BOURLET, JAMES, *Economic Research Council*
BRIERLY, D. J., *Shropshire County Council*
BROOKS, P., *Tilling Management Services*
BURGESS, TYRELL, *North East London Polytechnic*
BUTLER, P., *Department of the Environment*

CARDONA, GEORGE, *The Treasury*
CAVE, MERRIE, *Hammersmith and West London College*
CLARK, PETER, *BBC*
COOK, CLAUDIA, *United Newspapers*
CRAWLEY, JOHN, *Central Policy Review*

DAVENPORT, IAN, *Western Heritable Investment Co.*
DENISON, E. A. K., *North Yorkshire County Council*
DRIVER, R. K., *Kent County Council (Education)*

EGERTON, JOSEPH, *ABCC*
ELLIOTT, CLIVE, *IEA Subscriber*
EYRES, STEPHEN, *Free Nation*

FAWKNER, JOHN
FLEW, A., *University of Reading*
FROST, G., *Centre for Policy Studies*

GERRARD, CECELIA, *Surrey Education Committee*
GOULD, J., *University of Nottingham*
GRUGEON, SIR JOHN, *Kent County Council*

HARDMAN, SIR HENRY
HARTLEY, J.P., *CBI*
HARTLEY, KEITH, *University of York*
HATHERLEY, JOHN, *Kings College School, Wimbledon*

HILDRETH, JAN
HOPPÉ, MALCOLM, *Aims*

JACKSON, DEREK, *Freedom Association*
JACKSON, W. U., *Kent County Council*
JONES, ARTHUR
JONES, GEORGE W., *London School of Economics*
JONES, ROB, *Federation of Civil Engineering Contractors*

KALETSKI, ANATOL, *Financial Times*
KERR, ANDREW
KING, DAVID, *University of Stirling*

LAIT, JUNE, *University College, Swansea (University of Wales)*
LAMB, RICHARD, *Journalist*
LEECH, ANDREW, *BWS Publishing Ltd*
LEWIS, RUSSELL, *Daily Mail*
LICHFIELD, N., *University of London/Nathaniel Lichfield & Partners*

MARGOLIS, CECIL, *Cecil Margolis (Harrogate) Ltd*
MARSLAND, DAVID, *Brunel University*
MEDDINGS, F., *Retired Banker*
MICKLETHWAIT, BRIAN, *Adam Smith Society*
MORGAN, J., *Association of District Councils*
MOWAT, DAVID, *Edinburgh Chamber of Commerce*

PESCHEK, DAVID, *Local Government Information Service*
PREST, ALAN, *London School of Economics*

QUORROLL, B. J., *Slough Borough Council*

RANTALA, LAURIE, *London School of Economics*
REDDELL, A., *Department of the Environment*
REED, C. T.
REID, WILLIAM, *University of Birmingham*
RICHARDS, GORDON, *Hammersmith and West London College*
RITCHIE, RICHARD, *Selsdon Group*
ROBSON, BARBARA

SANDLER, HECTOR, *University of Mexico*
SAWERS, DAVID
SHARROCK, W. W., *University of Manchester*
SHERMAN, ALFRED, *Centre for Policy Studies*
SQUIRE, ROBIN, *MP*
STAFFORD, DAVID, *University of Exeter/Deputy Leader, Exeter City Council*
STERN, M., *IEA Subscriber*
STUTCHBURY, O., *Dartington & Co.*
SUMBY, ALAN, *GLC*

THOMAS, A. G.
THONEMANN, P. C., *University College, Swansea*
TRAVERS, H. A.
TUNLEY, P., *Higher Education Review*
TYNDALL, JOHN, *Journalist*

WALDERN, JOANNE, *CBI*
WALSH, KIERON, *University of Birmingham*
WEBB, A. J., *CBI*
WELLS, ALAN J., *London Borough of Ealing*
WHETSTONE, FRANCIS
WHETSTONE, LINDA
WOODS, J. K., *Lacsab*

IEA READINGS in Print: Recent Issues

16. The State of Taxation
A. R. Prest, Colin Clark, Walter Elkan, Charles K. Rowley, Barry Bracewell-Milnes, Ivor F. Pearce; *and eleven commentators with an Address by* Lord Houghton
1977 (xvi+116pp., £2·00)

18. The Economics of Politics
James M. Buchanan, Charles K. Rowley, Albert Breton, Jack Wiseman, Bruno Frey, A. T. Peacock, *and seven other contributors.*
Introduction by Jo Grimond
1978 (xiii+194pp., £3·00)

19. City Lights
Essays on financial institutions and markets in the City of London
E. Victor Morgan, R. A. Brealey, B. S. Yamey, Paul Bareau
1979 (x+70pp., £1·50)

20. Job 'Creation'—or Destruction?
Six essays on the effects of government intervention in the labour market
John Addison, Christian Watrin, Malcolm R. Fisher, Albert Rees, Yukihide Okano and Mitsuake Okabe, Walter Eltis *with an Introductory Essay by* Ralph Harris
1979 (xiii+146pp., £3·00)

21. The Taming of Government
Micro/macro disciplines on Whitehall and Town Hall
Stephen C. Littlechild, Gordon Tullock, A. P. L. Minford, Arthur Seldon, Alan Budd, Charles K. Rowley *with an Introduction by* Lord Robbins
1979 (xxiii+136pp., £3·00)

22. Tax Avoision
The economic, legal and moral inter-relationships between avoidance and evasion
A. R. Ilersic, Anthony Christopher, D. R. Myddelton, Christie Davies, Lord Houghton; with a Prologue by Arthur Seldon and an Epilogue by Barry Bracewell-Milnes
1979 (x+134pp., £2·50)

23. The Prime Mover of Progress
The Entrepreneur in Capitalism and Socialism
Israel Kirzner, Leslie Hannah, Neil McKendrick, Nigel Vinson, Keith Wickenden, Sir Arthur Knight, Sir Frank McFadzean, P. D. Henderson, D. G. MacRae, Ivor F. Pearce
1980 (xiv+154pp., £3·50)

24. Is Monetarism Enough?
Essays in refining and reinforcing the monetary cure for inflation
Patrick Minford, Harold Rose, Walter Eltis, Morris Perlman, John Burton; *and ten other contributors.*
1980 (x+118pp., £3·00)